AGRICULTURAL METEOROLOGY

THE AUTHORS

Dr. Vishal Dattatray Wale Completed Bachelor's degree (B.Sc. Agriculture), Master's degree (M.Sc. Agriculture) in Agricultural Meteorology, and a Doctoral degree (Ph.D.) in Agricultural Meteorology from Mahatma Phule Krishi Vidyapeeth, Rahuri, Maharashtra. He has qualified the National Eligibility Test conducted by ASRB. He has written 11 research papers in highly NAAS rating journals, books, and a number of popular and technical articles periodically in popular magazines. He has 6 years of professional experience in the field of agricultural meteorology. He has actively participated in many conferences and trainings programs and is a member of the Association of Agrometeorologists.

Dr. Sachin Yadavrao Wankhede Completed Bachelor's degree (B.Sc. Agriculture) from Dr. Panjabrao Deshmukh Krishi Vidyapeeth, Akola (MH) and Master's degree (M.Sc. Agriculture) in Agricultural Meteorology and a Doctoral degree (Ph.D.) in Agricultural Meteorology from Mahatma Phule Krishi Vidyapeeth, Rahuri (MH). He has 8 years of professional experience in the field of agricultural meteorology, and he has written 15 research papers in highly NAAS rated journals, and authored 3 abstracts, 3 book chapters, and a number of popular and technical articles. He has actively participated in scientific exhibitions, international and national training, national worksop and webinars/seminars. Currently, working as a Subject Matter Specialist in agrometeorology at KVK, ICAR-Central Institute for cotton research, Nagpur.

Dr. Shraddha Vikrant Bagade Completed Bachelor's degree (B.Sc. Horticulture), Master's degree (M.Sc. Agriculture) in Agricultural Meteorology and a Doctoral degree (Ph.D.) in Agricultural Meteorology from Mahatma Phule Krishi Vidyapeeth, Rahuri (MH). She has 21 years of experience and has written 40 scientific papers, 07 technical papers, 08 book chapters, 06 books/bulletins and a number of popular articles. She has attended international training, National workshop, national training, long-term and short-term training and conducted 2 training programmes as course co-ordinator. Currently, working as an I/C, Head & Assistant Professor of Agricultural Meteorology, College of Agriculture, Pune.

Dr. Jayawant Dadaji Jadhav Completed Bachelor's degree (B.Sc. Agriculture) and Master's degree (M.Sc. Agriculture) in Agricultural Meteorology from Mahatma Phule Krishi Vidyapeeth, Rahuri, Maharashtra and, a Doctoral degree (Ph.D.) in Agricultural Meteorology from Gujarat Agricultural University, Anand. He has 35 years of experience in research and agricultural extensities and has written more than two hundred scientific publications, more than a hundred technical and extension publications and radio/TV talks, 32 bulletins and 22 book chapters. He has many national and state level awards. He is the Ex. Head of Department of Agricultural Meteorology, College of Agriculture, Pune.

AGRICULTURAL METEOROLOGY

– Authors –

Vishal Dattatray Wale

Sachin Yadavrao Wankhede

Shraddha Vikrant Bagade

Jayawant Dadaji Jadhav

2026

Daya Publishing House®

A Division of

Astral International Pvt. Ltd.

New Delhi – 110 002

ISBN 978-93-5919-693-0

Published by : **Daya Publishing House®**
A Division of
Astral International Pvt. Ltd.
– ISO 9001: 2015 Certified Company –
4736/23, Ground Floor, Ansari Road,
Darya Ganj, New Delhi-110 002
Ph. 011-43549197, 8130496929
e-mail: info@astralint.com
Website: www.astralint.com

Printed at : **Replika Press Pvt Ltd.**

Preface

Agriculture plays an important role in addressing livelihood, poverty, hunger, malnutrition and employment of millions of peoples in India and in the word. Agriculture exhibits a fluctuating trend in recent years may due to climate variability and climate change and there by changing trends of monsoon patterns. Climate and weather are the most important in agriculture. Climate of a region determines the nature of crops to be grown and weather during crop periods determines the yield potentials. A critical knowledge on weather and climate and their direct and indirect effects on crop growth and development are essential for successful crop production.

Agricultural meteorology is the study of weather and climate in relation to agriculture, has become increasingly vital as global environmental changes continue to shape the landscape of farming and food production. With the growing pressures of climate change, water scarcity, and increasing food demand, understanding the intricate relationship between weather, climate, and agricultural practices is more critical than ever.

In a world where unpredictable weather patterns and extreme events are becoming more frequent, farmers and agronomists must rely on the insights provided by agricultural meteorology to make informed decisions and adapt to changing conditions. This book aims to provide a comprehensive overview of agricultural meteorology, presenting both theoretical foundations and practical applications of primitive aspects of meteorology and its application in the field of agriculture.

The authors earnest hope is that the book would prove valuable for students and teachers in the field of agricultural meteorology. Suggestions and comments from students, academicians and scientists towards the improvement of the text would be most welcome and may be communicated to the author.

Vishal Dattatray Wale
Sachin Yadavrao Wankhede
Shraddha Vikrant Bagade
Jayawant Dadaji Jadhav

Contents

Chapter 1

Introduction to the Sun, the Earth and the Atmosphere

1.1 The Sun

The Sun is the star at the center of the solar system. Its diameter is about 109 times that of earth and its mass is about 330,000 times that of earth, accounts for about 99.86 per cent of the total mass of the solar system, about 75 per cent of the solar mass consists of hydrogen, 23 per cent helium and less than 2 per cent consists of heavier elements, including carbon, nitrogen, oxygen, neon, magnesium, silicon and iron. The surface temperature of sun is approximate **6000°K.**

The sun consists of three parts; photosphere, chromospheres and corona. The photosphere is an outer visible layer of the sun with a temperature of 6400°K. The chromospheres are a thin layer of the sun roughly, 2,000 km just above the photosphere with a temperature of 50000°K. The corona is a region around the sun, extending from chromospheres with a temperature of 2 MK and it emits X-ray radiations. The Sun has several layers beneath the visible surface; the core, radiative zone and convective. The core is approximately 20 per cent of the size of the solar interior and is thought to have a temperature of approximately 15 MK, making it the hottest part of the Sun. The light generated in the core is transported through the radiative zone very slowly, since the high density of matter in this region means a photon cannot travel too far without encountering a particle, causing it to change direction and lose some energy.

The convective zone is the outermost layer of the solar interior. It is a thick layer that transports energy from the edge of the radiative zone to the surface through giant convection cells. Sunspot is region on sun surface appears darker than surrounding because of lower temperature. Solar flare is a sudden brightening over the sun surface. The constant flow of particles from the sun is known as solar wind and are composed of high energy protons.

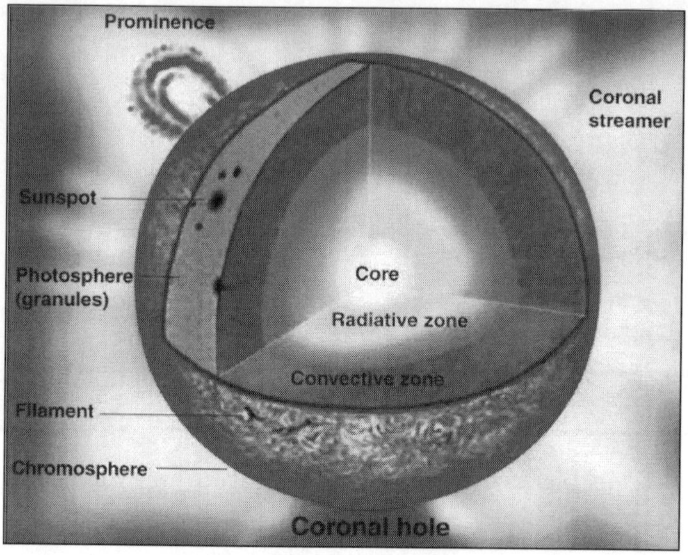

Figure 1.1. Structure of the Sun.

The mean distance of the sun from the earth is approximate 149.6 × 10^6 km (1,49,597,890 km) [1AU (Astronomical Unit), it is a unit of length in Astrophysics] *i.e* 8 minute and 19 second at light speed; the maximum distance of the sun from the earth is approximate 152.1 × 10^6 km and the minimum distance of the sun from the earth is approximate 147.1 × 10^6 km. The electromagnetic radiation emitted by the sun is our primary source of energy. The energy is due to internal hydrogen-helium fusion reactions. All matter present in the sun is in the form of gas and plasma because of its high temperature. This makes it possible for the sun to rotate faster at its equator (about 25 days) than it does at higher latitudes (about 35 days near its poles).

1.2 The Earth

It does not have its own source of illumination. Its structure is oblate ellipsoid. The mass of earth is 5.97 × 10^{24} kg, surface pressure is 101.325 kPa and mean surface temperature is 15°C (285°K). Earth's systems are divided into four spheres that make up the world; air (atmosphere), water (hydrosphere), land (lithosphere) and living organisms (biosphere).

The earth has two types of motions;

a) Rotation

It rotates west to east upon its axis. It completes one rotation in 24 hrs. The time period of earth rotation determines the length of a day. The rotational speed is maximum at equator and minimum at poles. The rotation not only causes days and night but also determines direction of winds and ocean currents.

Figure 1.2. Different Spheres of the Earth.

b) Revolution

The rotating earth revolves on elliptical orbit around the Sun. The time taken to complete one revolution around the sun is called a year.

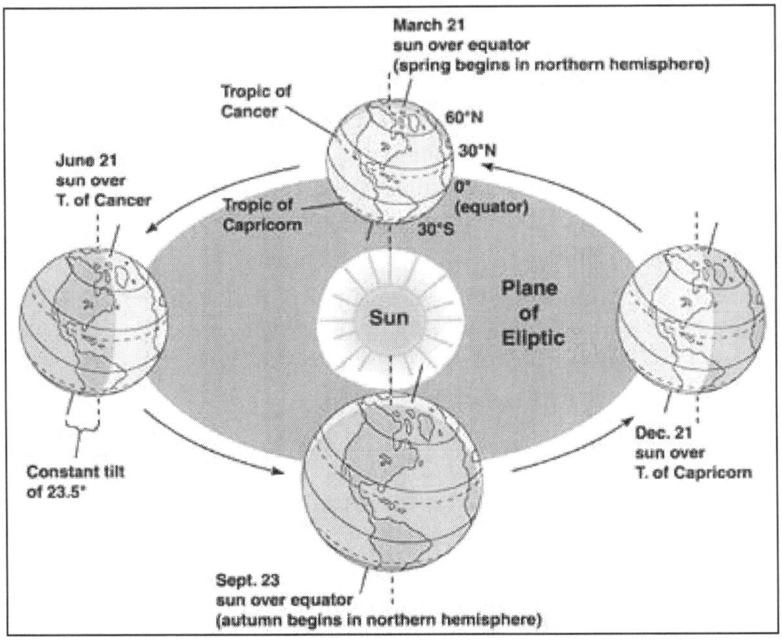

Figure 1.3. The Effect of Earth's Rotation and Revolution.

The time period of earth revolution determines the length of a season. The earth complete one revolution in 365.25 days. Hence, leap year comes after four years in February month. The axis of the earth rotation inclined about 66.5° from the plane of the elliptic. An imaginary plane passed through the sun and extended outward through all points in the earth's orbit is called plane of the elliptic.

1.3 Distances between the Sun and the Earth

The longest distance between the sun and the earth during earth revolution is 152.5 million km is known as Aphelion and it is observed in 4th July. The nearest distance is 147.5 million km is known as Perihelion and is observed in 3rd January. The average distance between sun and earth is about 149.6 million km.

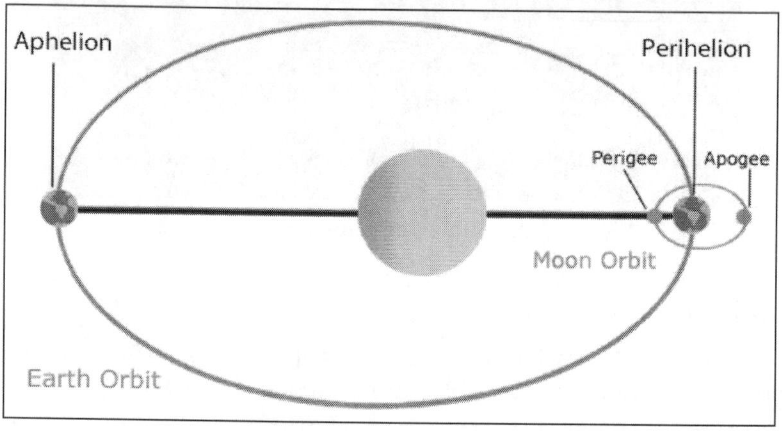

Figure 1.4. Points between Sun and Earth.

Solar day is a time required for a single rotation of the earth on axis with respect to the sun. Polar day is the region of the earth when the day or sun stay above the horizon for more than 24 hours. Polar night is the region of the earth when the night lasts for more than 24 hours. Subsolar point is the place where the sun lies directly overhead at noon.

Perigee is the least distance between the moon and the earth. Apogee is the greatest distance between the moon and the earth. The distances can be expressed in terms of speed of light which is about 3×10^8 ms^{-1}. At that speed, light from the Sun reaches the earth in 8.33 minutes.

1.4 Equinox

The word equinox derived from **Latin** words *aequinoctium, aequus* means equal and *nox* means night. When the sun rays are over head or vertical end of equator (00 Latitude), this situation comes **twice** in a year during earth revolution. When sun rays on the equator then the day and night duration is same throughout the globe.

a) Vernal Equinox

The point at which sun crosses from south to north pole on the equator is called vernal equinox and it is observed in 21st March in the northern hemisphere. In the northern hemisphere the March equinox is known as the vernal equinox, and in the southern hemisphere as the autumnal equinox.

b) Autumnal Equinox

The point at which sun crosses from north to south pole on the equator is called autumnal equinox and it is observed in 23rd September northern hemisphere. In the northern hemisphere the September equinox is known as the autumnal equinox, and in the southern hemisphere as the vernal equinox.

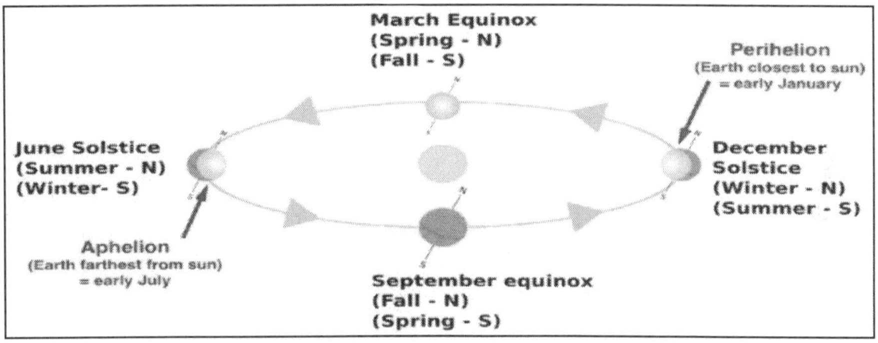

Figure 1.5. Equinox and Solstice.

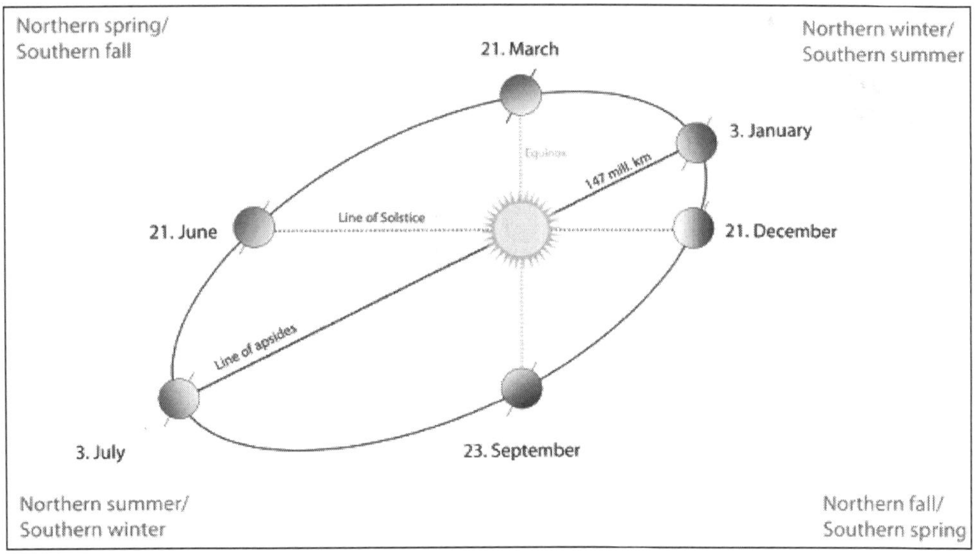

Figure 1.6. Revolution of the Earth.

1.5 Solstice

The word solstice derived from Latin words, *Sol* means sun and *sistere* means to stand still. The time during revolution of the earth at which it reaches its maximum declination, when days and nights are unequal is called solstice.

A) Summer Solstice

When day is longest and night is shortest and it is observed 21st June. The sun rays are vertical at tropic of cancer. The tropic of cancer is a line of latitude circling the earth at 23.5° N of the equator.

B) Winter Solstice

When night is longest and day is shortest and it is observed in 22nd December. The sun rays are over head at tropic of Capricorn. The tropic of capricorn is a line of latitude circling the earth at 23.5° S of the equator.

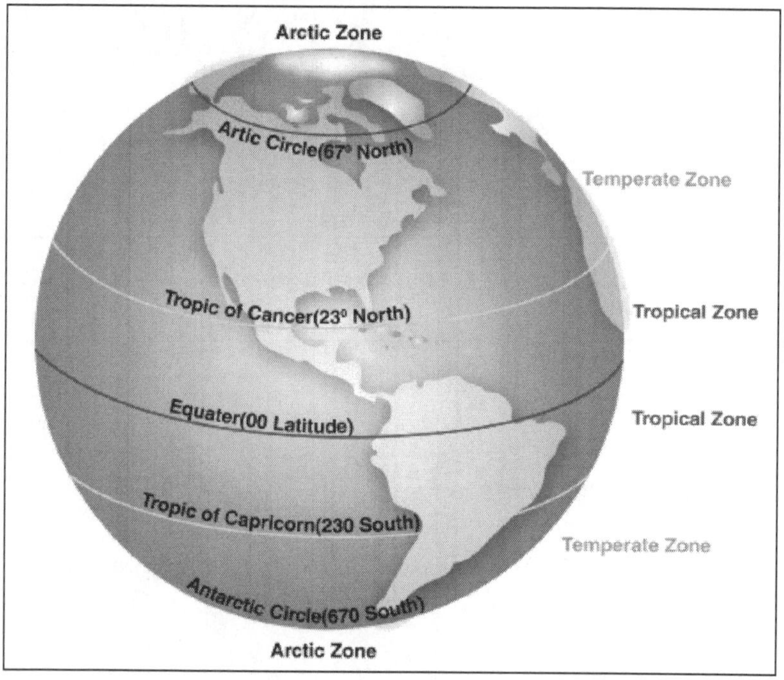

Figure 1.7. Latitude of the Earth Systems.

1.6 Atmosphere

The word atmosphere derived from Greek word *atmos* means vapour and *sphaira* means sphere. Atmosphere is the gaseous envelop surrounding the earth surface or Atmosphere is a colourless, tasteless and odourless mixture of gases that surround the earth surface. The height of atmosphere is about 1600 km, Mass is about 56 × 10^{14} tonnes (99 per cent of atmosphere mass is within 40 km from the earth and 50 per cent of atmosphere mass is within 5.5 km from the earth surface), mean sea level atmospheric pressure is 1013.25 mb and temperature is 15°C. The earth temperature in absence of atmosphere would have been + 95°C in day and at night -145°C.

1.6.1 Importance of Atmosphere

- It provides the oxygen for the survival of living organisms.
- It provides carbon dioxides to build in photosynthesis.
- It protects biological life from harmful UV rays.
- It supplies necessary precipitation or moisture.
- It acts as a medium for transport of pollens, seeds, *etc.*
- It shields the planet from harmful solar radiation.
- It regulates the earth's temperature through greenhouse effect, crating condition suitable for life.
- It provides nitrogen, which is essential for plant growth.
- It facilitates the water cycle by enabling evaporation, condensation and precipitation.
- It fulfills the biological oxygen demand of animal life. [BOD is the amount of oxygen consumed by bacteria and other microorganisms while they decompose organic matter under aerobic (oxygen is present) conditions at a specified temperature.]

1.6.2 Composition of Atmosphere

The various components of the atmosphere can be divided into; gases, water vapour and aerosol.

a) Gases

The gases of the atmosphere remain constant at surface of globe up to the height of 80 to 88 km known as non-variable or permanent gases. They are nitrogen (78.08 per cent), oxygen (20.95 per cent), argon (0.93 per cent), neon (0.0018 per cent), helium (0.0005 per cent), krypton (0.00011 per cent) and xenon (0.00009 per cent). the gases of the atmosphere changes with change in time, space, season *etc.* is known as variable gases. They are carbon dioxide (0.03 per cent), carbon monoxide, nitrous oxide (0.00005 per cent), sulphur dioxide, ozone (0.00004 per cent), hydrogen (0.00006 per cent), methane (0.00017 per cent) and ammonia (trace).

b) Water Vapour

Moisture in the form of water vapour exist in the atmosphere. It is released through the process of evaporation and transpiration. It is also important components of hydrological cycle. Water vapour comprise up to 4 per cent of the atmosphere by volume and 3 per cent by weight near the surface, but it is almost absent above 10 to 12 km. Water vapour is a gas that varies significantly from time to time and place to place (variable gas).

c) Aerosol

The solid particles (dust, salt, carbon) and liquid particles (water droplets) present in the air known as aerosols. Bacteria, spores, pollen, seeds, smokes are also present in the atmosphere. They are also called as aerosol. Aerosol enter the atmosphere by manmade pollution, agricultural practices and forest fire.

1.6.3 Stratification or Physical Structure of Atmosphere

1.6.3.1 Structure of Atmosphere on the Basis of Chemical Composition

a) Homosphere

The layer at which chemical composition of gas is uniform is called as homosphere. This is due to transportation of gases, diffusion of gases, turbulent mixing and convection. The altitude of this sphere is nearly 80-88 km.

b) Heterosphere

The layer at which chemical composition of gas is not uniform is called as heterosphere. The distribution of gases is governed by the gravitational force. Thus, heavier gases sink downward while the lighter gases at highest altitude;

Nitrogen and Oxygen layer: upto 88-115 km

Atomic Oxygen layer: upto 115-965 km

Helium layer: upto 965-2400 km

Hydrogen layer: upto 2400-10000 km

1.6.3.2 Structure of Atmosphere on the Basis of Vertical Temperature Variations

A) Lower Atmosphere

a) Troposphere

The word troposphere derived from Greek word *tropos* means mixing or turbulence. It is the lower layer of the atmosphere. It extends to an elevation of 16-18 km over equator and 8-10 km over poles above the earth surface. This layer contains almost all the water and dust and represent all the atmospheric phenomena such as clouds, rain, snow, fog, lightning, thunder storms, cyclone *etc.* No visible weather phenomenon occurs above this layer. So, this layer is called as seat of weather phenomena. Decrease of temperature with increasing height at a mean lapse rate of about 6.5 °C/km or 3.6°F/1000 feet. This phenomenon is called as lapse rate. Wind velocities increase with height above the earth surface. It contains 75 per cent of the gaseous mass of the total atmosphere. Temperature reaches up to -50 to -60°C at tropopause. Atmospheric pressure changes from

1013.25 to 101.13 mb. Tropopause is an isothermal layer above the troposphere which separate stratosphere from troposphere.

The atmospheric boundary layer is the lowest part of troposphere that is directly influenced by the presences of the earth's surface and responds to surface forcing within a timescale of about an hour or less. It is the region of the lower troposphere where the Earth's surface strongly influences temperature, moisture, and wind through the turbulent transfer of air mass. It is also known as Peplosphere or Ekman layer or Planetary boundary layer. Ekman layer was named by the V.W. Ekman, vertical region of the ocean affected by the movement of wind-driven surface waters. The Planetary boundary layer exists up to a height of 1000 meters.

b) *Stratosphere*

The elevation of this layer is about 50 km above the earth surface. The temperature increases with increasing height is known as temperature inversion. It contains much of the total atmospheric ozone. The density of ozone is maximum at 22 km. Hence this layer is called as Seat of photochemical reactions. Temperature reaches upto 0°C at above 50 km because ozone absorbs the ultraviolet rays from sun. The air in this sphere is very dry, *i.e.* less than 0.01 gm of water vapour/kg of air. Atmospheric pressure changes from 101.325 mb to 1 mb. Stratopause is an isothermal layer above the stratosphere which separate mesosphere from stratosphere.

B) *Upper Atmosphere*

c) *Mesosphere*

The elevation of this layer is about 80 km above the earth surface. Temperature decreases with increasing height (Lapse rate). Temperature reaches upto -90°C at mesopause. In this layer the temperature decreases with height at the rate of 5°C/km. It is the coldest region in the atmosphere. Atmospheric pressure is very low and decreases from 1 mb to 0.01 mb. Mesospause is an isothermal layer above the mesosphere.

d) *Thermosphere*

It is the uppermost layer of atmosphere; elevation is about 400 km. Temperature increases with increasing height (temperature inversion). Temperature reaches to 950°C to 1700°C. Above 80 km ionization occurs due to solar X-ray and UV radiation. Ionosphere, Exosphere layer present above the thermosphere.

Karman Line

It is the altitude where, space begins. It commonly represents the border between the Earth's atmosphere and outer space. It is 100 km (about 62 miles) high. The *exobase* is the level in the atmosphere above which the mean free path of particles is longer than a scale height.

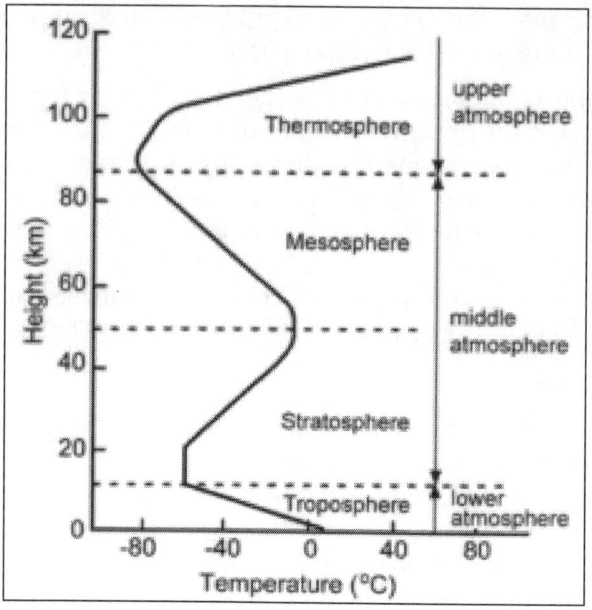

Figure 1.8. Temperature based Vertical Layers of Atmosphere.

It is the outermost region of the earth's atmosphere, lying above the thermosphere and extending thousands of kilometers into space, from which molecules having sufficient velocity can escape the earth's gravitation.

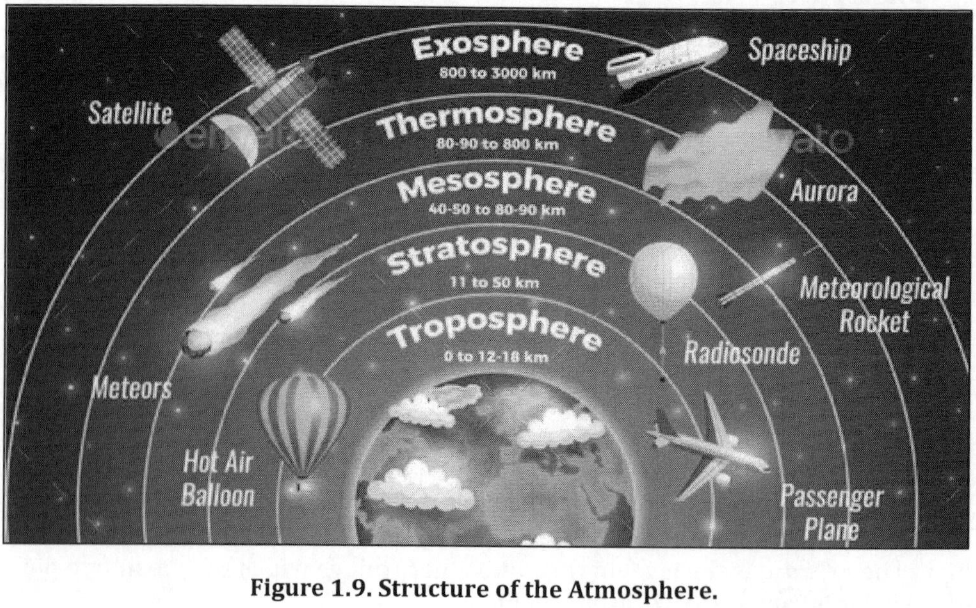

Figure 1.9. Structure of the Atmosphere.

Meteoroid

It is a small rocky or metallic body in outer space. When a meteoroid, enters in earth's atmosphere, aerodynamic heating of that object produces a streak of light, both from the glowing object and the trail of glowing particles that it leaves in its wake. This phenomenon is called a meteor or shooting star.

Radiosonde

It is a small, expendable instrument package that is suspended below a balloon filled with hydrogen or helium. As radiosonde rises at about 1,000 ft/min (300 m/min), sensors on the radiosonde measure profiles of pressure, temperature and relative humidity.

C) Other Sphere

a) Ozonosphere

The lower part of stratosphere between 15-35 km is concentrated with ozone (O_3) and is called ozonosphere. The highest density is as about 22-25 km height. Ozone absorbs ultraviolet radiation and protects biological life from harmful radiation. Therefore, ozonosphere is called as umbrella of the earth. The ozone layer was discovered in 1913 by the French physicists Charles Fabry and Henri Buisson. The ozone layer absorbs 97-99 per cent of the sun's medium frequency ultraviolet light (from 200 nm to 315 nm wave length). The ozone layer can be depleted by free radical catalysts, including nitric oxide (NO), Nitrous oxide (N_2O), Hydroxyl (OH), CFCs (carbo fluorocarbon), BFCs (Bromo fluorocarbon) *etc.*

b) Ionosphere

It is also known as Kennelly Heaviside layer and elevation is about 100-300 km. It is important for transmission of radio waves at longer distance. The density of electrons is nearly 5-10 electron/cm^3.

Auroro

Due to extraction, light is emitted and the sky during night become luminous; Aurora borealis in northern hemisphere and aurora australis in southern hemisphere. Air glow which originates with self-luminescent gases. Auroras are the result of disturbances in the magnetosphere caused by solar wind.

c) Exosphere

The elevation of this layer is about 500-750 km. Hydrogen and Helium gases predominate in this outer most region. Helium particles are produced by the action of cosmic radiation on nitrogen and from the decay of radioactive elements in the earth crust.

d) Magnetosphere

The elevation of this layer is about 2000 km. Only electron and proton are present in this layer. Distribution of particles is governed by magnetic field of the earth and not by gravity.

G) Van Allen Belts

This layer is also known as Van Allen radiation belts and elevation is about 4000 to 20000 km. In this layer charge particles are concentrated.

Chapter 2

Introduction to the Agricultural Meteorology

2.1 Meteorology

The word meteorology had been derived from Greek word *meteoros* means above the earth's surface or lower atmosphere and *logus* means study or science. The word meteorology was given by Aristotle and defined meteorology as a study of lower atmosphere. There are many definitions of meteorology few are given here; It is the science dealing with physics of atmosphere. It is a branch of physics of the earth dealing with physical processes in the atmosphere that produce weather. It is science of atmosphere and its phenomenon weather and climate).

Branches of Meteorology

a) Synoptic Meteorology

It studies the relationship between the atmospheric circulations and the surface environment of a region. It is to prepare large scale wind system over a region and give weather forecast on region wise. It concerns with the relationship between atmospheric circulation and local climate.

b) Dynamic Meteorology

It explains the motions and energy transformation, which occurs in the atmosphere, using the tools of mathematics and physics. It is an objective method of weather forecasting, also known as Numerical Weather Prediction.

c) Physical Meteorology

It is the branch of meteorology dealing with the study of optical, electrical, acoustical, and thermodynamic phenomena in the atmosphere, including the physics of clouds and precipitation.

d) Statistical Meteorology

It is the branch of meteorology deals with the mean state of the physical properties of the air. This branch is popularly known as climatology.

e) Agricultural Meteorology

It is the science, which deals with the physical characteristics of the environment of growing plants or animal organisms.

2.2 Agricultural Meteorology

J. W. Smith defined agricultural meteorology as meteorology in its relation to agriculture. Prasada Rao defined agricultural meteorology as the science which deals with the influence of weather on crop husbandry and animal husbandry. Agricultural meteorology is an applied science which deals with relationship between weather or climatic conditions and agricultural production. Agrometeorology is a science investigating the meteorological, climatological and hydrological conditions, which are significant for agriculture owing to their interaction with the objects and processes of agricultural production.

2.2.1 Scope of Agricultural Meteorology

- ☆ **Characterization of agricultural climate for determining crop growing season:** The elements of weather are important factors on which the growth, development and yield of a crop depends. On the basis of climatic condition, agricultural year can be divided into three seasons *i.e. kharif, rabi* and summer. Agro meteorology consider and assesses the suitability of these parameters in a given region for maximum crop production and economic benefits.
- ☆ **Crop planning for stability in production:** Proper knowledge about weather elements helps in crop planning and stabilized yield even under adverse climatic conditions. Stability in production can be increased by selecting short duration crop, low water requirements crop, drought resistant or tolerant crops. Contingent crop planning can also be done for stability in production.
- ☆ **Crop management on weather basis:** Calendar of operation can be carried out on the basis of weather support and weather forecasts from agromet advisories. *e.g.* fertilizer application is advised when rainfall is less than 1 mm/day, wind speed is less than 30 km/hr, soil moisture is between 30-80 per cent. Spraying or dusting is undertaken when there

is no rainfall, adequate soil moisture and wind speed is less than 25 km/hr.

- **Crop monitoring:** The monitoring of crop growth and performance during developmental stages is an important aspect of agricultural management. To check crop health and growth suitable meteorological tools such as remote sensing, crop simulation model, crop growth models, water balance technique can be used.
- **Crop modelling:** A model is a mathematical equation giving growth or yield as a function of climatic parameters. A crop simulation model is a simulation model that describes processes of crop growth and development as a function of weather conditions, soil conditions and crop management.
- **Useful in response farming:** The prediction of monsoon season based on analysis of weather date of pre-monsoon period and managing crops accordingly.
- **Soil formation and development:** The soil formation processes; physical, chemical and biological depends on climatic factors like temperature, precipitation, humidity, pressure, wind *etc.*
- **Yield forecasting:** This is useful to help collecting firms planning their grain harvest and storage and to support the decisions of companies investing in regional agricultural activities. Crop yield forecasting is important for national food security, including early determination of the import or export plan and price, providing timely information for optimum management of growing crops.
- **Research in crop climate relationship:** Agrometeorology can help to understand crop climate relationship so as to resolve complexities of plant process in relation to its micro climate. It also develops suitable technique of modification of micro climate.
- **Forecasting climatic extremities and their management:** Climatic extremities such as frost, floods, drought, high winds, hailstorms can be forecasted and the loss in production can be prevented by arranging suitable strategies.
- **Diagnose soil moisture stress:** Climatic water balance method is used to determine soil moisture and soil moisture stress, from that drought and protective measures such as irrigation, mulching, application of antitransparent, defoliation, thinning can be undertaken.
- **Livestock production:** The favorable and unfavorable weather conditions for growth, development and production of livestock is studied in agrometeorology. Thus, to optimize milk production and poultry production the climatic normal are worked out.

- **Soil formation:** Soil formation process depends on climatic factors like temperature, precipitation, humidity, wind, pressure *etc.*
- **Remote sensing in agriculture for crop monitoring:** It is used in crop acreage estimation, crop monitoring in relation to moisture and nutrient stress, disease and pest attack, crop health and yield forecast.
- **Agro meteorological services:** Agro meteorological services release weather bulletins to help the farmers. *e.g.* The Agro-meteorological Advisory Service (AAS) rendered by IMD, to apply relevant meteorological information to help the farmer make the most efficient use of natural resources, with the aim of improving agricultural production; both in quantity and quality. The agro met advisory bulletins are issued in twice in a week *i.e.*, Tuesday and Friday at district, state and national levels.
- **Modification of microclimate:** Agro-meteorology studies important to develope techniques of modification of microclimate to increase yield, assists the breeders in developing tailor made varieties and to develope suitable instruments or sensors to record plant responses.
- **Minimize field trials:** Agro meteorology providing mathematical basis helps to minimize field trials for crop management.

2.2.2 Importance of Agricultural Meteorology in Crop Production

- It helps in planning of cropping patterns and cropping systems.
- It is used to develop weather based agro advisories to sustain crop production.
- It helps in weather based effective farm operations.
- It helps to prepare crop weather diagram and crop weather calendar.
- It reduces the loss of applied agrochemicals and fertilizers.
- It helps managing weather abnormalities *i.e.* cyclones, heavy rainfall, floods, drought by lifesaving irrigation, avoid fertilizer application and chemical sprays, use shelter belts *etc.*
- It reduces the loss of applied agrochemicals and fertilizers.
- It is used to monitor agricultural droughts on crop-wise for effective drought management.
- It helps in selection of sowing dates for optimum crop yields.
- It helps in cost effective calendar of operation *i.e.* ploughing, harrowing, weeding *etc.*
- It is useful in efficient irrigation scheduling to crops.
- It is useful in efficiently and timely harvesting of all crops.
- It reduces or eliminates outbreak of pests and diseases.
- It is effective for environmental protection.
- It avoids or minimize losses due to forest fires.

2.1.3 Different Between Meteorology and Agricultural Meteorology

Meteorology	*Agricultural Meteorology*
It is branch of atmospheric physics	It is branch of applied meteorology
It is a weather science	It is a product of agriculture and meteorology
It is a physical science	It is a biophysical science
It aims at weather forecasting	It aims at improving quantity and quality of crop production through meteorological skills
Weather service is the concern	Agro-advisory service to the farmers is the concern based on weather forecast
It is a linking science to the society	It is a linking science to the farming community

2.3 Historological Events in Meteorology and Agricultural Meteorology

384-322 BD	Aristotle has written first book on meteorology entitled "Meteorologica".
1450	Leone Battista Alberti developed a first swinging plate anemometer.
1593	Galileo discovered thermometer to enable measurement of air temperature.
1636	Halley published a treatise on Indian summer monsoon and attribute its seasonal reversal of wind due to differential heating of land masses.
1639	Benedetto Castelli discovered first rain gauge in Italy.
1643	Toricelli discovered barometer to enable measurement of atmospheric pressure.
1643	Blaise Pascal and Rene Descartes demonstrated that atmospheric pressure decreases with increasing altitude using a barometer.
1645	Boyle showed that barometer can be used for weather prediction.
1648	Blaise Pascal rediscovers that atmospheric pressure decreases with height.
1662	Christopher Wren invented mechanical, self-emptying tipping bucket rain gauge.
1667	Robert Hooke invented a swing type anemometer for measuring wind speed.
1676	Ole Roemer is the first person to measure the speed of light.

1686	Edmand Halley explained trade and monsoon winds and published climatological map.
1714	Gabriel Fahrenheit creates a reliable scale for measuring temperature with mercury type thermometer.
1730	Priestly discovered the oxygen.
1735	George Hadley explained effect of earth rotation on direction of trade.
1735	Reaumur quantify the relationship between the amount of heat and plant development.
1735	George Hadley ideal explanation of global circulation through study of the trade winds.
1742	Anders Celsius proposed degree Celsius (°C) temperature scale.
1749	Wilson studied kite experiments for upper air temperature variation.
1752	Franklin showed electricity on the cloud is similar to the electricity produced by friction in lab.
1759	M. V. Lomonsore recognized practical importance of meteorology to agriculture (Russia).
1761	Joseph Black discovers that ice absorbs heat without changing its temperature when melting.
1772	Daniel Rutherford discovers nitrogen.
1777	Antoine Lavoisier discovers oxygen.
1780	Horace Desaussure invented the hair hygrometer for measuring humidity.
1783	De Saussure used human hair to measure the moisture of the air and constructed the hair hygrometer.
1784	Asiatic Society of Bengal was Founded to promote scientific studies in meteorology.
1785	First meteorological observatories were established at Calcutta to study weather and climate in India.
1787	Jacques Charles discovered the relation between temperature and volume of air.
1793	Meteorological observatory was established in Madras.
1800-1850	Dove explained general circulation of the earth (Germany).
1800-1850	Redfield explained origin of the cyclone (USA).
1800	Jean Nicholas Fortin invented a portable mercury barometer.
1801	Lamark was made first cloud classification.

1803	Luke Howard devised a classification of cloud that was adopted by the IMO in 1929.
1805	Francis Beaufort designed wind scale to estimate velocity and strength of the wind.
1806	Francis Beaufort introduced system for classifying wind speeds.
1824	J. B. Fourter proposed greenhouse effect.
1825	August was devised Psychrometer.
1826	Meteorological observatory was established in Colaba.
1827	H. W. Dove was developed the laws of storm.
1835	Gaspard Coriolis demonstrated the effect that the earth's rotation has on atmospheric motion.
1835	Hendry Piddington published papers on tropical storm in Journal of Asiatic Society in Calcutta and coined term cyclone.
1842	Hendry Piddington in Calcutta published law of storm.
1844	Lucien Vidi invented the aneroid barometer.
1844	Marconi started wireless communication for collecting upper air data.
1845	Berghans was prepared first world map of precipitation.
1845	Thomas Robinson invented simple type of anemometer.
1848	Baron Kelvin prosed temperature scale °K.
1853	John Campbell invented Campbell Sunshine Recorder and modified in 1879 by George Gabriel Stokes.
1854	Reutovich published first book on Agro meteorology (Russia).
1856	William Ferrell proposed the existence of a circulation cell in the mid latitudes.
1857	Buys Ballot and William Ferrel was showed that surface wind flows from high to low pressure and is deflected by a force called Coriolis force, caused by the earth's rotation around its own axis.
1872	Robert Angus Smith coined the term acid rain.
1873	International Meteorological Organization (IMO) was established.
1875	Imperial (India) Meteorological Department (IMD) was established at Alipore, Calcutta.
1889	Sir John Eliot was appointed as first director general of observatories.
1896	Svante Arrhenius proposed a relation between atmospheric CO_2 concentration and temperature.
1900	De Candole first time classified the climate based on vegetation.
1905	IMD head office shifted from Calcutta to Shimla.
1913	Charle Fabry and Henri Bussion discovered ozone layer.

1916	J. W. Smith published book on Agricultural Meteorology (USA).
1909	Sir Gilbert Walker developed multiple regression model for forecasting monsoon rainfall over India.
1919	Gardner and Allard discovered photoperiodism, photo periodic induction (USA).
1920	Bowen with Bowen ratio (The ratio of sensible to latent heat fluxes at the surface).
1920	Bjerknes explained polar front theory of formation of extra tropical cyclone.
1923	Gilbert Walker proposed the walker cell model and Southern Oscillation.
1926	John Patterson developed three cup anemometer.
1928	IMD head office shifted from Shimla to Pune.
1928	Hopkin was given bioclimatic law.
1932	The division of Agricultural Meteorology was started and it became a permanent department of India Meteorological Department in 1943.
1932	Dr. L. A. Ramdas shows horizontal and vertical climatic variation near ground surface.
1935	Lysenko postulated that plants response differs to physical environment during various stages of growth.
1936	Koppen classify the climate based on weather elements.
1944	S. K. Banerjee was first Indian Director of IMD.
1945	Farmer weather bulletins was started by IMD.
1946	Vincent Schafer discovered the principle of cloud seeding.
1948	Penman and Thornthwaite discovered the concept of PET.
1948	IMD and ICAR had launched a crop weather scheme for compilation of phenological data and weather.
1949	IMD became a member of WMO.
1950	Evelyn Pruitt coined the term remote sensing.
1950	Rainfall Atlas of India was published.
1951	Swisbank developed the techniques to estimate exchange of momentum and energy fluxes in the soil plant atmosphere continuum (SPAC).
1951	World Meteorological Organization was started in Geneva, Switzerland.

1952	F.M. Missorie developed phytotromes (Artificial climatic chamber) and studied the responses of the plants to artificially controlled climatic condition.
1959	Monteith shows light distribution and radiation reflection by vegetation.
1960	Establishment of Climatology section at CAZRI, Jodhpur.
1960	Fischer coined the term Remote Sensing.
1960	Launch of the first successful weather satellite (TIROS-1).
1962	Indian Institute of Tropical Meteorology (IITM) was established.
1970	Doorenboss and Pruitt shows crop water requirement.
1970	The research in agro-meteorology was initiated on crop wise for the first time India at ICRISAT.
1970	National Commission on Agriculture.
1972	Space application center, Ahmadabad was started.
1974	Rosenberg study of microclimate among soil, plant and atmosphere to understand SPAC system.
1975	The IMD is initiated AAS in collaboration with State Agricultural Department.
1976	Agro-meteorological advisories were first initiated to provide state level forecast.
1976	The National Commission on Agriculture has recommended a separate Department of Agro-meteorology in all SAU and ICAR research institute to strengthen teaching, research and extension in Agricultural meteorology and to train the technical personnel.
1983	AICRP on Agro-meteorology was started by ICAR under CRIDA.
1988	NCMRWF was established.
1990	Agro ecological region map of India was prepared.
1999	Agro ecological sub-region map of India was prepared.
2004	National Commission on Farmer.
2006	IARI was developed InfoCrop crop simulation model.
2007	Integrated Agro-meteorological Advisory Service was launched.
2008	United Nation general assembly declared the year as the International Year of Planet Earth.
2008	District Agro-meteorological Advisory Service was launched.
2008	National Remote Sensing Centre, Hyderabad.
2010	SAFAR (System of air quality, weather forecasting and research) system was developed by IITM, Pune with IMD and NCMRWF to provide location specific information on air quality.

2011	National Innovation on Climate Resilient Agriculture was started to develop dynamical prediction system for monsoon rainfall.
2012	National Monsoon Mission.
2012	Gramin Krishi Mausam Sewa was launched by IAAS.
2016	IMD launched System of Aerosol Monitoring and Research (SAMAR) to study the concentration of Black carbon, radiative properties of aerosols, environmental visibility and their climatological impacts.

2.4 Important Organization in the Field of Agricultural Meteorology

2.4.1 World Meteorological Organization

Figure 2.1. World Meteorological Organization.

It is a specialized agency of the United Nations responsible for promoting international cooperation on atmospheric science, climatology, hydrology and geophysics. The WMO originated from the (IMO) International Meteorological Organization a non-governmental organization founded in 1873 as a forum for exchanging weather data and research. The WMO was established by the Convention of the World Meteorological Organization, signed 11th October 1947 and ratified on 23rd March 1950 and Headquartered located Geneva, Switzerland. The WMO is made up of 193 countries and territories, and facilitates the "free and unrestricted" exchange of data, information, and research between the respective meteorological and hydrological institutions of its members. It has six regional specialized meteorological centres of tropical cyclone in New Delhi (India), Nadi (Fiji), Miami (US), Honolulu (Hawaiian), Tokyo (Japan) and La Reunion (France). It also collaborates with nongovernmental partners and other international organizations on matters related to environmental protection, climate change, resource management, and socioeconomic development. World meteorological day is celebrated every year on 23rd March.

2.4.2 India Meteorological Department

India Meteorological Department was established in 1875 and headquarter located at New Delhi. It is an Agency of the Ministry of Earth Sciences of the Government of India. It is the principal agency responsible for meteorological observations, weather forecasting and seismology. It became a member of the World Meteorological Organization after independence on 27th April 1949. It plays a vital role in preparing the annual monsoon forecast, as well as in tracking the progress of the monsoon across India every season. In collaboration with the

Indian Space Research Organisation, the IMD also uses the IRS series and the Indian National Satellite System (INSAT) for weather monitoring of the Indian subcontinent. It is one of the regional Specialised Meteorological Centres of the Tropical Cyclone Programme of the World Weather Watch of WMO. It acts as RSMC, New Delhi to provide tropical cyclone and storm surge guidance to 13 member countries under WMO/ESCA panel. It is regional nodal agency for forecasting, naming and disseminating warnings about tropical cyclone in the Indian Ocean north of the Equator. It acts as tropical cyclone aviation to Asia pacific middle east countries for civil aviation. It has three tier networks with cyclone warning division at New Delhi, 3 area cyclone warning centre (Mumbai, Chennai, Kolkata), 4 cyclone warning centre (Bhubaneshwar, Vishakhapatnam, Ahmedabad, Tiruvanathapuram). It has Regional meteorological centres at New Delhi, Chennai, Mumbai, Guwahati, Kolkatta, Nagpur. It has Meteorological centres at Agartala, Amaravati, Bhopal, Chandigarh, Gangtok, Hyderabad, Bengaluru, Bhubaneshwar, Lucknow, Raipur, Shillong, Srinagar, Ahmedabad, Jaipur, Dehradun, Goa, Itanagar, Leh, Patna, Ranchi, Shimla, Tiruvanathapuram).

Figure 2.2. India Meteorological Department.

2.4.3 IITM

The Indian Institute of Tropical Meteorology (IITM) was established on 17th November 1962 as Institute of Tropical Meteorology at Pune as a distinct unit of India Meteorological Department for the study of the fundamental atmospheric problems in the tropical region. The Institute was transferred into an autonomous organization on 1st April 1971 under the name Indian Institute of Tropical Meteorology. It studies on weather forecasting, climatology, monsoon studies, climate modeling, hydrometeorology, weather modification, atmospheric chemistry, atmospheric electricity, cloud physics and instrumentation for observational studies. The IITM has been playing important role in the international activities such as Inter Governmental Panel on Climate Change (IPCC), International Geosphere Biosphere Programme (IGBP), Indian Climate Research Programme (ICRP).

Figure 2.3. Indian Institute of Tropical Meteorology.

2.4.4 NCMRWF

The National Center for Medium Range Weather Forecasting (NCMRWF) is a centre of excellence in weather and climate modelling under Ministry of Erth Sciences, Government of India was established in 1988 with the mandate to provide medium range (3-10 days in advance) location specific forecasts to farmers through development of Agro-Advisory Service for the country. The Centre has implemented and upgraded from time to time a global Numerical Weather Prediction (NWP) system for preparation and dissemination of medium range weather forecasts. This requires the use of voluminous global atmospheric and oceanographic observations, their assimilation to provide initial conditions, numerical models, and high-end computing resources. The Centre established a network of Agro-Meteorological Field Units in the Agro-climatic zones of the country through which agro advisories were issued to farming community. After the successful demonstration of Agro-advisory services in real time, the same has been handed over to India Meteorological Department for the operational use.

Figure 2.4 National Center for Medium Range Weather Forecasting.

2.5 Weather and Climate

2.5.1 Weather

It is the instantaneous physical state of the atmosphere at particular place and time. It is a condition of the atmosphere at a given place and at a given time.

Elements of Weather

Any entity that controls the weather is called as element of weather. There are various factors which individually or in combination control the weather, which are; solar radiation, temperature, air pressure, wind, sunshine, humidity, cloudiness, precipitation, *etc.* All these elements are highly variable and constitute the weather. Day to day or hour to hour changes in weather are mainly the result of variations in the amount, intensity and distribution of weather elements over earth.

a) Solar Radiation

Sun is the nearest star to the planet earth and is the primary sources of energy. Solar energy provides light required for seed germination, leaf expansion, growth of stem and shoot, flowering, fruiting and thermal conditions necessary for the physiological functions of the plant. Solar radiation plays an important role as regulator and controller of growth and development. Solar radiation also influences assimilation of nutrient and dry matter distribution. Light is the visible portion of the solar spectrum with wavelength range is from 0.4 to 0.7 nm. Light is essential for the synthesis of the chlorophyll pigment which absorbs the radiant energy and converts it into potential energy of carbohydrate. The carbohydrate, thus formed is the connecting link between solar energy and living world. In addition, it regulates the important physiological functions. The characteristics of light *viz.* intensity, quality, duration and direction are important for crops.

b) *Temperature*

It is the degree of hotness or coldness. Solar radiation is the main source for air temperature. It is one of the important variables, which influences all stages of crop during its growth development and reproductive phase. It is useful in leaf production, expansion and flowering, growth and development, different biochemical reactions in crops.

c) *Humidity*

It is closely related to rainfall, wind and temperature. It plays a significant role in crop production. It determines the crops grown in a given region. It affects the internal water potential of plants, physiological phenomena, transpiration, potential evapo-transpiration, water requirement, irrigation water requirement. It enhances the growth of some saprophytic and parasitic fungi, bacteria and pests, the growth of which causes extensive damage to crop plants.

d) *Precipitation*

Precipitation is a major component of the water cycle and is responsible for depositing the fresh water on the planet. Precipitation helps farmers to grow crops and provides a fresh water supply for us to drink. The monsoon accounts for over 70 per cent of the country's annual rainfall and has an enormous impact on the Indian economy. The quantity of rainfall determines agricultural production. Time, amount, frequency, form and extent of precipitation influences crop production. On the basis of water requirement of plants, the plants are grouped into following groups; hydrophytes are the plants adapted to aquatic environment. *e.g.* floating rice, xerophytes are the plants adapted to grow in dry lands where water content is low. *e.g.* cactus and mesophytes are the plants living in the habitat that usually shows neither an excess or nor a deficiency of water. *e.g.* wheat, maize.

e) *Pressure*

The pressure exerted by the atmosphere on the earth's surface is called atmospheric pressure. Atmosphere contains air molecules which have a certain weight, it exerts pressure on the earth surface. Generally, in area of higher temperature, atmospheric pressure is low and in areas of low temperature, pressure is high. With increase in elevation, the atmospheric pressure decreases. Atmospheric pressure does not have influence on crop growth. It is however an important parameter in weather forecasting.

f) Wind

Wind is air in horizontal motion which travels from a high-pressure area to a low-pressure area. Wind direction and velocity have significant influence on crop growth. Wind increases the turbulence in atmosphere, thus increasing the supply of carbon dioxide to the plants resulting in greater photosynthesis rates, increases transpiration especially cuticular transpiration than stomatal transpiration, hot wind accelerates the drying of the plants by replacing humid air by dry air in the inter cellular spaces. Strong wind damages the shoots, lodging (paddy, sugarcane, banana *etc.*), flower and fruit shedding, crops and trees with shallow roots are uprooted, cold wind causes chilling injuries, causes soil erosion and soil deposition causes poor aeration in root zone.

g) Clouds

Solar radiation provides energy for evaporation. Evaporation supplies water vapour to the air. Air rises upwards on account of increasing temperature. As the mass of air goes up, it expands due to low pressure and cool. If the cooling proceeds to saturation, water vapour condenses and cloud formation takes place. The clouds are aggregation of minute drops of water suspended in the air at higher altitudes.

2.5.2 Difference Between Weather and Climate

Weather	*Climate*
It is an instantaneous physical state of atmosphere	It is a generalized physical state of atmosphere
It is expressed in numerical values	It is expressed in terms of average values
Weather refers to short term duration	Climate refers to long term changes
It changes from place to place at the same time	It requires longer period for changes
It provides meteorological information	It provides a geographical information
It is measured in observatory	It is measured by series of observation
It is categorized as fair, unfair	It is classified as tropical, arid, marine
It decides the success or failure of a crop in a growing season	It decides the suitability of introduction of new crop in a region
Weather on micro scale can be modified	It is difficult to modify
It is useful in short term contingent planning	It is useful in long term agricultural planning
It does not include magnitude of extreme values of weather elements	It includes extremes of weather elements. *e.g.* cyclone, flood, frost, drought, hailstorm

2.5.3 Climate

The word climate comes from the Greek word *'klima'* means slope. Climate is an average state of weather. or Climate is the average of weather information of a particular area during a specified period of time (more than 30 years). The branch that studies the climate called as climatology. Agricultural climatology is the branch of climatology that applies climatic knowledge to understand agricultural problems.

2.5.3.1 Classification of Climate

De Candole (1900 AD) is the first person to attempt the classification of climate based on vegetation. Koppen (1936) is first person to classify climate based on weather elements *i.e.* temperature and precipitation. Thornthwaite classify the climate based on precipitation of effectiveness index, temperature effectiveness index and seasonal distribution of rainfall. Troll (1965) classifies the climate based on temperature and humid month. All India coordinated Research Project on Dryland Agriculture of Indian Council of Agricultural Research adopted classification based on moisture deficit index.

I) Agro-climatic Classification Based on Rainfall

- ☆ Arid zone: The region with annual rainfall is less than 250 mm.
- ☆ Semi-arid zone: The region with annual rainfall is between 250 to 500 mm.
- ☆ Sub-humid zone: The region with annual rainfall is between 500 to 1000 mm.

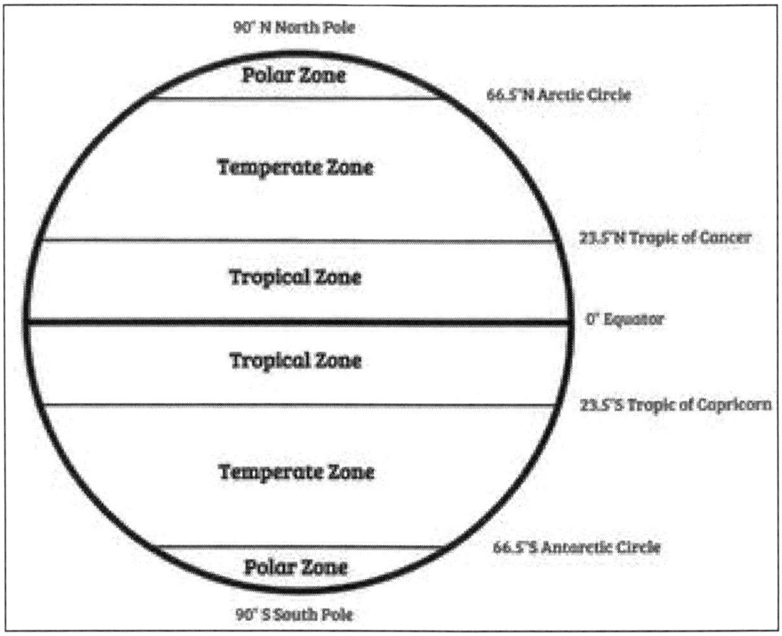

Figure 2.5. Classification of Climate Based on Latitude.

- Humid zone: The region with annual rainfall is between 1000 to 1500 mm.
- Wet Humid zone: The region with annual rainfall is more than 1500 mm.

II) Classification of Climate Based on Latitude

- **Tropical Climate:** The tropical climate zone stretches from the Tropic of Cancer at 23.5° north latitude to the Tropic of Capricorn at 23.5° south latitude, with the equator centered within this zone.
- **Subtropical Climate:** The subtropical climate are located between the tropical zone and the temperate zone (25° to 40° North and South latitude) in the northern and southern hemisphere.
- **Temperate Climate:** The temperate climate zone falls between 23.5° and 66.5° north and south latitudes. Extending from the southern edge of the Arctic Circle to the Tropic of Cancer in the northern hemisphere, and the northern edge of the Antarctic Circle to the Tropic of Capricorn in the southern hemisphere.
- **Polar Climate:** The polar climate zones fill the areas within the Arctic and Antarctic circles, extending from 66.5° north and south latitude to the poles.

III) Classification of Climate by Koppen's

Koppens has taken annual and monthly means of temperature and precipitation for the classification of climate.

Depending on temperature, the climate is divided into;

- Tropical climate: The coldest month average temperature is more than 18°C.
- Temperate climate: The coldest month average temperatures in between 18°C to -3°C.
- Cold climate: The coldest month average temperatures is less than -3°C.
- Polar climate: The average temperature of the warmest month is less than 10°C.

Depending on precipitation, the climate is divided into:

- Rainy climate: Precipitation is received in the form of rain or liquid form.
- Snow climate: Precipitation is received in the form of snow or solid form.

Based on these parameters, Koppen classified climate into five groups;

- Tropical rainy climate: The coldest month average temperature is more than 18°C, precipitation is received in the form of rain and rainfall is more than evaporation.

- ☆ Tropical dry climate: The coldest month average temperature is more than 18°C, rainfall is less than evaporation.
- ☆ Mild temperate rainy climate: The coldest month average temperatures is in between 18°C to -3°C.
- ☆ Cold snow climate: The coldest month average temperatures is less than -3°C and precipitation is received in the form of snow.
- ☆ Polar climate: The warmest month average temperatures is less than 10°C.

IV) *Classification of Climate by Thornthwaite*

Thornthwaite classified climate based on PE (precipitation of effectiveness) index. TE (temperature effective) index and seasonal distribution of rainfall.

$$\text{PE index} = \frac{\textit{Annual precipitation}}{\textit{Annual precipitation}} \times 100$$

$$\text{TE index} = \sum_{i=1}^{12} \frac{T\,(oF)-32}{4}$$

a) *Based on PE Index*

Symbol	*Humidity Province*	*PE Index*
A	Wet	128 and above
B	Humid	64-127
C	Sub-humid	32-63
D	Semi-arid	16-31
E	Arid	Below 16

b) *Based on Seasonal Distribution of Rainfall*

Symbol	*Seasonal distribution*
R	Rainfall adequate in all seasons
S	Rainfall deficit in summer
W	Rainfall deficit in winter
D	Rainfall Deficit in all seasons

c) *Based on TE index*

Symbol	*Temperature Province*	*TE Index*
A	Tropical	128 and above
B	Meso thermal	64-127

Symbol	*Temperature Province*	*TE Index*
C	Micro thermal	32-63
D	Taiga	16-31
E	Tundra	1-15
F	Frost	0

V) Classification of Climate by Trolls

Troll defines humid month, as the month having precipitation more than the Potential evapo-transpiration (PET). Troll took temperature and humid months as parameters for distinguishing different climate. based on the climate data, ICRISAT revised trolls' classification and classified climatic regions of India in arid climate, semiarid dry climate, semiarid wet climate and humid climate.

Sl.No.	*Climate*	*No. of Humid Month*
1	Arid	< 2
2	Semi-Arid	2.0 to 4.5
3	Wet	4.5 to 7.0
4	Humid	> 7.0

VI) Classification of Climate by Moisture Deficit Index

All India Coordinated Research Project on Dry land Agriculture of Indian Council of Agricultural Research adopted classification based on Moisture Deficit Index (MDI)

$$MDI = \frac{P - PET}{PET} \times 100$$

where,

P = Annual precipitation

PET = Potential evapotranspiration

Based on MDI, the climate is divided into three regions as;

Type of Climate	*MDI*
Per humid	+100 and above
Humid	+20 to +100
Moist sub humid	0 to +20
Dry sub humid	-33.33 to 0
Semiarid	-66.66 to -33.33
Arid	-100 to -66.66

2.6 Climatology

The study of distribution of climatic elements over a region for a long period of time is known as climatology. Simply climatology is the study of climate. Climatology is the combination of two Greek words, *klima* and logos, *klima* means slope of earth and logos means study. Climatology is the science dealings with the factors which determine and control the distribution of climate over the earth's surface.

2.6.1 Branches of Climatology

a) **Physical climatology** is the branch of climatology which seeks to describe the factors responsible for bringing about the temporal and spatial variations in heat, moisture and momentum exchange between earth surface and its atmosphere and within atmosphere. Observations on climatic elements help to study physical environment. These climatic elements, individually or in combination are the resultant of complex exchange processes of heat, moisture and momentum.

b) **Regional climatology** is the branch of climatology, which determines and describes the different types of climates. On the basis of spatial scale climatic region is divided into macro-climatic region, meso climatic region and micro-climatic region. Continental or oceanic factors control macroclimate whereas meso climates and micro climates are influenced by local factors.

c) **Microclimatology** is the branch of science which study climatic elements in a particular location *e.g.* in field crops, orchards, animal shed *etc.*

d) **Applied climatology** is the branch of climatology, which study the application of climatological knowledge to specific practical problems.

e) **Agricultural climatology** is the science, which applies climatic knowledge to understand agricultural problems, or it may be defined as study of climatic elements in relation to agriculture. Agroclimatology deals with the relationship of climatic regimes and agricultural production.

2.7 Seasons and Climate

The season form due to revolution of the earth around the Sun. In the northern hemisphere the season are as follows;

- ☆ Spring: March, April and May
- ☆ Summer: June, July and August,
- ☆ Autumn: September, October and November
- ☆ Winter: December, January and February

For agriculture purposes, the year has been divided into three crop seasons as;

- ☆ *Kharif* season: 15 June to 15 October
- ☆ *Rabi* season: 15 October to 15 February
- ☆ Summer season: 15 February to 15 June.

The IMD has divided the climate year into 4 season;

- ✰ Summer (March to May)
- ✰ Monsoon (June to September)
- ✰ Post monsoon (October to November) and
- ✰ Winter (December to February)

2.8 Hopkins Bio Climatic Law

Hopkins attempted to express the importance of latitude, longitude and altitude in the distribution and rate of development of plants. It shows a lag of four days for each degree of latitude, five degrees of longitude and 400 m of altitude, northward, eastward and upward in spring and early summer. The sequence of flowering obeys Hopkins Bioclimatic law. The time of flowering develops upon the latitude, longitude and altitude.

- ✰ For every degree of latitude (north or south to equator), flowering is retarded by 4 calendar days.
- ✰ For every 5° of longitude for (East to west) on land areas flowering is advanced by 4 calendar days.
- ✰ For each 400 m increases in altitude flowering is retarded by 4 calendar days.

2.9 Length of Growing Periods (LGP)

It is the duration of rainy season and the period for which the soil moisture storage at the end of post rainy season. LGP is a period in which the available soil moisture is enough to meet the Evapotranspiration requirement of the crop. The LGP is the number of days in year during which rainfall and moisture stored in the soil exceeds half of potential evapotranspiration. The spatial distribution of crops and farming system in any region is determined by the LGP. Information on LGP helps in selection of cultivars of a particular crop. The length of the rainy season is the duration on between the onset and end of agricultural significant rains. The LGP is the length of the rainy season.

Length of Growing Period	*Climate*	*Cropping System*	*Rainfall*
< 75 days	Arid climate	Mon cropping	< 750 mm/annum
75 – 140 days	Semi-arid climate	Sole cropping	
140 – 180 days	Humid climate	Inter cropping	> 750 mm/annum
> 180 days	Per humid climate	Double cropping	> 900 mm/annum
		Multiple cropping	> 1200 mm/ annum

Chapter 3

Atmospheric Pressure

3.1 Introduction

Pressure can be defined as the force (F) per unit surface area (A). Air pressure is a force exerted by mass of a column of air above a given point. Thus, atmospheric pressure can be defined as the weight exerted by air column on unit surface of the earth. Atmosphere pressure is measured by an instrument called as barometer. There are four types of barometers; fortin's barometer, kew pattern barometer, aneroid barometer and barograph.

3.2 Units of Measurement

a) Height of mercury column measured in inch, cm or mm.

b) 1 Bar = 1000 mb

= 10^6 dynes/cm^2

c) Pascal (1 pa = 1 N/m^2 = 10 dyes/cm^2) is a standard international (SI) unit of pressure.

1 atm = 29.92 inches

= 76 cm or 760 mm of Hg

= 1013 milli bar (mb)

= 101325 pascals

= 101.325 kilo pascal (KPa)

= 1.014 × 106 dynes/cm^2

3.3 Variation of Atmospheric Pressure

3.3.1 Vertical Variation of Atmospheric Pressure

The air pressure depends on the density of the air and density of air depends on its temperature, composition and force of gravity. The density of air decreases with increase in height so that pressure also decrease with increase in height. The atmospheric pressure decreases rapidly on an average at the rate of about 34 mb/300m height or the mean rate of decrease in pressure with height is 1cm/100m. In the first few thousand meters above the sea level, the pressure decreases at the rate of 1mb/10m.

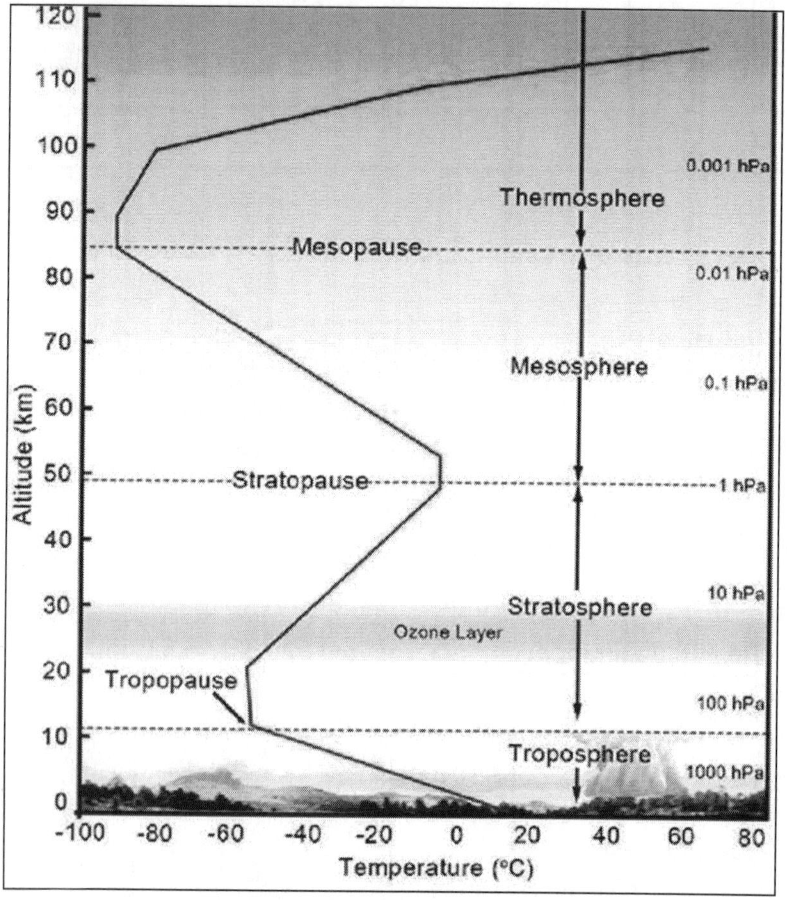

Figure 3.1. Vertical Variation of Atmospheric Pressure.

The atmospheric pressure at sea level is 1013.25 mb, at 11-16 km it is about 101.325 mb, at 50 km it is about 1 mb, and at about 80 km it decreases to 0.01 mb. The rate of fall of pressure with height is determined partly by rate of fall in temperature.

3.3.2 Horizontal Variation of Atmospheric Pressure

The horizontal distribution of air pressure mainly depends upon; pressure gradient, coriolis, frictional and centrifugal forces.

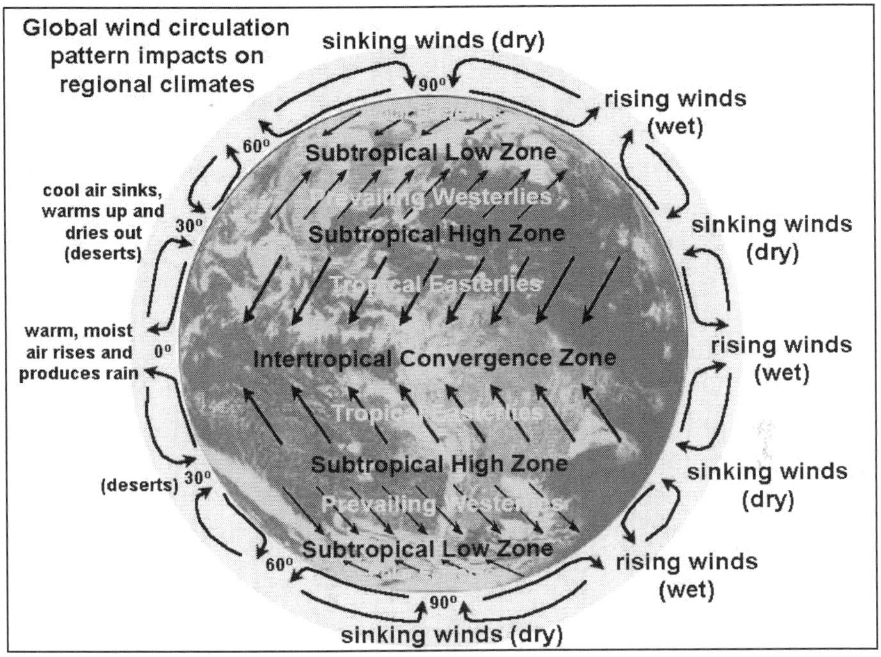

Figure 3.2. Horizontal Variation of Atmospheric Pressure.

The distribution of atmospheric pressure across the latitude is termed as global horizontal distribution. This distribution is characterized by presence of distinctly identifiable zones of homogeneous pressure regime or pressure belts. Globe can be divided into following pressure belts; Equatorial low-pressure belt is low-pressure belt exists over equatorial region between 5°N and 5°S and is due to thermally produced high insolation over the equatorial region. Sub-tropical high-pressure belt is pressure belt lies between 25° and 35° N and S in both hemispheres. Sub polar low-pressure belt is belt of low pressure lies between 60° and 70° latitude in both the hemispheres and belt results due to rotational movement of earth. Polar high-pressure belt is pressures belt at poles are consistently high throughout the year. The surface atmospheric pressure changes cause for both the monsoon *i.e.* southwest monsoon and northeast monsoon during summer and winter. Isobar is a line connecting various places, having equal atmospheric pressure. The differences in atmospheric pressure with horizontal distances is known as the pressure gradient. Issallobar is a line connecting places of equal atmospheric pressure gradient. The atmospheric pressure is high during winter while low during summer, followed by post monsoon season. The low atmospheric pressure during summer leads to monsoonal wind over India. The atmospheric pressure is high during the morning hours and later on decreases with temperature increases.

3.4 Factors Affecting Pressure Distribution Over the Earth Surface

- ☆ **Altitude:** Air pressure decreases with altitude. The places at higher altitudes have lower air pressure than the places at lower altitudes.
- ☆ **Latitude:** The air temperatures are high over equatorial region and low over polar regions. Therefore, air pressure is low at equator and high over poles. The air pressure increases move towards poles.
- ☆ **Land and water surface:** Air pressure is more over water surface as compared to land surface in summer season and vice-versa in winter season. Air pressure is higher over water surface in comparison with nearby land surface during day time and vice-versa during night.
- ☆ **Ocean currents:** Air pressure will be lower over the places near warm ocean currents in comparison with the places near cold ocean currents.
- ☆ **Condition of atmosphere:** Presence of dust particles in atmosphere increases its temperature, thereby air pressure decreases. If water vapour content increases in air, it becomes less dense as compared to dry air. Therefore, air pressure decreases with increase in water vapour content of the air at same temperature.

3.5 Important Facts about Atmospheric Pressure

- ☆ **Isobar** is the line joining the places of equal atmospheric pressure on the weather map.
- ☆ **Trough** is the V-shaped projection of low pressure (valley of low pressure).
- ☆ **Low pressure area** is the low-pressure at center enclosed with circular closed isobars.
- ☆ **Pressure gradient** or **isobaric slope** is the rate of change of pressure between two places at same elevation.
- ☆ **Depression** is a low-pressure area with wind velocity less than 39 miles/hr or 34 knots.
- ☆ **Secondary depression** is a small depression formed on the outside of a parent depression.
- ☆ **Super cyclone** is the cyclone in which wind speed more than 220 kmhr.
- ☆ **Ridge** is wedging shaped extension of high pressure which lies between two depressions.
- ☆ **Col** is pressure distribution between two high pressure areas and two low pressure areas.
- ☆ **Vortex** is a mass of air or water that spins around very fast and pulls objects into its empty center.

3.6 Cyclone and Anticyclone

3.6.1 Cyclone

The Harry Piddington published paper on tropical storm and coined the term cyclone. The term cyclone is derived from the Greek word *"cyclos"* that means "Coils of Snake". It is a rotating low pressure weather system. It is a circular shape of isobars having low pressure at center and pressure increases outwards. The direction of rotation of cyclone in the northern hemisphere is anticlockwise, while in the southern hemisphere it is clockwise. The pressure gradient force and Coriolis force cause air flow in cyclones to spiraling convergent system. Cyclones are also known as lows. Examples: Fani, Nargis, Amphan, Phyan, Nisarg, Tauktae *etc.*

Characteristics of Cyclones

- ☆ **Pressure distribution:** The lowest pressure is in the center and it increases towards the outer rim gradually.
- ☆ **Temperature distribution:** The rear and left-hand sides facing the direction of it are the coldest while the front and right-hand side are warmest.
- ☆ **Moisture distribution:** Relative humidity increase towards the center of the cyclone.
- ☆ **Cloud distribution:** Cirrus cloud appear in the front.
- ☆ **Rainfall distribution:** Highest rainfall occurs at the front or warmer side but cooler sides have less frequent rains.
- ☆ **Wind velocity and directions:** Wind velocity increases from outer rim to the center. The velocity of wind is more than 34 knots.
- ☆ **Eye of the cyclone:** The center of the cyclone is the region with calm air, no cloudiness, little horizontal motion and no precipitation.

3.6.1.2 Classification of Cyclones

Category	*Wind Speed in Knots*	*Wind Speed in km/hr*
Low pressure area	Less than 17	Less than 31
Depression	17 - 27	31 - 49
Deep Depression	28 - 33	50 - 61
Cyclonic Storm	34 - 47	62 - 88
Severe Cyclonic Storm	48 - 63	89 - 118
Very Severe Cyclonic Storm	64 - 119	119 - 221
Super Cyclonic Storm	120 and above	222 and above

Examples

Name of Cyclone	*Type of Cyclone*	*History*	*Wind Speed (kmph)*	*Area Affected*
Bijli	Cyclonic Storm	April 2009	75	Eastern India
Asani	Severe Cyclonic Storm	May 2022	100	Andhra Pradesh, Tamil Nadu, Karnataka, Andaman and Nicobar
Michaung	Severe Cyclonic Storm	December 2023	100	Tamil Nadu, Andhra Pradesh
Burevi	Severe Cyclonic Storm	November 2020	85	Tamil Nadu, Kerala
Nisarga	Severe Cyclonic Storm	June 2020	110	Maharashtra
Ockhi	Very Severe Cyclonic Storm	November 2017	155	Southern India
Nivar	Very Severe Cyclonic Storm	November 2020	120	Andhra Pradesh, Tamil Nadu, Puducherry
Tauktae	Very Severe Cyclonic Storm	May 2021	185	Dadra, Nagar Haveli, Daman and Diu, Goa, Gujarat, Karnataka, Kerala, Lakshadweep, Maharashtra
Fani	Very Severe Cyclonic Storm	April 2019	215	Odisha West Bengal, Andhra Pradesh, East India
Gonu	Super Cyclonic Storm	June 2007	240	Western India
Amphan	Super Cyclonic Storm	May 2020	240	Odisha West Bengal, Andaman

3.6.1.3 Tropical Cyclone

It originates due to differential heating of land and water surface near islands in tropical region. A tropical cyclone is developed between tropic of Capricorn and tropic of cancer. The energy to the tropical cyclone comes from the latent heat released by the warm moist air. Hence, the force of the cyclone decreases with increase in distance from the sea. For the formation of tropical cyclone, the conditions required are; large sea areas for continuous supply of abundant warm and moist air, sea surface temperature more than 27°C, high Coriolis force (it makes the air to bend), preexisting low-pressure system and upper air divergence above the sea level. Tropical cyclones are absent in the belt of 5° latitude because of the absence of Coriolis force. A cyclone storm is a large vortex in the atmosphere with the diameter varies from 150 to 200 km. It is more violent and associated with very high winds (75-125 mile/hour) and storms producing rough weather. Tropical cyclones move with trade winds and most prominent in Pacific Ocean. They are also named as Hurricanes in USA or in Caribbean Sea, Typhoons in China and Willy-willies in Australia, Tropical cyclone in Indian Ocean, Baguio in Philippines and Taifu in Japan.

A mature tropical cyclone consists of region called as eye. An eye has a diameter of 40 km and characterized by weak wind shear, high temperatures, low pressure and cloud free. Around eye the wind speed is very high known as wall cloud region of cyclone storm. In this region thick cloud with embedded thunder storm extend about 15-16 km height. The wall cloud region produces heavy rainfall. This is the most dangerous part of the cyclone. Surrounding the wall cloud region wind is continues to be spiral in decreasing rate. The Tropical Cyclone Warning Centre is located at Jakarta, Indonesia. In India, IMD is nodal agency and responsible for meteorological observation. A cyclone is predicted by Area Cyclone Warning Centers (ACWC), Kolkata and Chennai in Bay of Bengal and Area Cyclone warning Center, Mumbai in Arabian seas, Cyclone Warning Center at Vishakhapatnam, Bhubaneswar and Ahmadabad. Both are sending their report to National Cyclone Warning Center (NCWC). The cyclone warning process is coordinated by the weather central in the office of DDGM (Weather forecasting) at Pune and the Northern Hemispheric Analysis center at New Delhi.

3.6.1.4 Temperate Cyclone

It is also known as wave cyclone or extra tropical cyclone or mid latitude cyclones. This type of cyclone originates in temperate or cold regions or in middle latitudes between 35° to 65° in both hemispheres. Its intensity is less compared to tropical cyclone but have more spread of several hundred kilometers. Its vertical thickness is 9-11 km and horizontal spread is of about 1000 km. It is associated with cloudiness, light rains and winds. They are formed due to convergence of two contrasting air masses. *i.e.*, warm, moist tropical air masses and cold, dense polar air masses. Polar fronts are created which are responsible for origin and development of temperate cyclone. They are capable of producing mild showers, heavy gales and tornadoes.

In 1920, V. Bjerknes and J. Bjerknes put fourth polar front theory of formation of extratropical cyclone based on the formation of fronts. The fronts are formed due to convergence of two air mass of different physical properties coming from different direction *i.e.* polar air mass (cold and denser) and tropical air mass (warm and moist) in northern hemisphere. When these two contrasting air masses converges along a line in mid latitude, they move parallel to each other, thus stationary front is formed. No cyclone develops over stationary fronts because there is no vertical movement of air. When two opposing air masses collide against each other and try to attack the territory of one another, unstable fronts are formed. In the beginning the surface separating two air masses is straight but it turns to unstable and wave like when warm air masses attempt to penetrate in the regions of one another. Such fronts are called as polar fronts. When warm and moist tropical air mass penetrates the territory of cold polar air mass along the polar front, it being lighter rises upward, with the result a center of low pressure is formed.

In an extratropical cyclone, the pressure at the earth surface may decrease by about 20 mb in 24 hrs. Now winds from all direction rush up towards the center of low pressure and thus cyclone is formed. Cyclone forming wave develops due to convergence (coming closer) of cold and warm air masses which divided into two groups (i) Eastern part, where eastward advancing warm tropical air mass ascends over cold air mass is called warm front (ii) In western part, where cold polar air mass pushes warm air mass upward is called cold front. Low pressure in the eastern part of cyclone is intensified with arrival of warm air. This draws the winds towards the center from the nearby areas. With the result cold front advances more rapidly than warm front. Consequently, cold and warm fronts come closer to each other resulting into the distraction of front. The cyclone dies due to disappearances of warm front, this process of cyclone destruction is called occlusion. The period of a cyclone *i.e.* cyclogenesis to frontogenesis or occlusion is called the life cycle of extratropical cyclone.

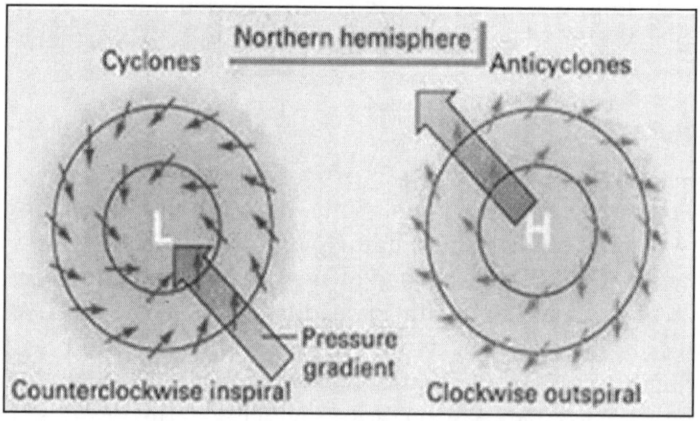

Figure 3.3. Cyclone and Anticyclone.

3.6.2 Anticyclone

It is a circular shape of isobars with high pressure at center and pressure decreases outwards. The direction of rotation of anticyclone in the northern hemisphere is clockwise, while in the southern hemisphere it is anticlockwise. These are also known as highs.

3.6.2.1 Characteristics of Anticyclones

- **Pressure distribution:** The highest pressure is in the center and it decreases towards the outer rim.
- **Temperature distribution:**Temperature near the ground is lower than the normal air temperature. At higher altitude temperature are generally higher.
- **Moisture distribution:** These are associated with dry, cool air, little cloud, slight precipitation. Air descent at the center, therefore temperature increase relative humidity and cloud dissipated.
- **Wind velocity and directions:** At ground air spirally rushes outwards from the center to periphery. Wind velocity is much lesser than in cyclones. Anticyclones move along with the general direction of the great air current.

Instruments Used to Measure Atmospheric Pressure

A) Barometer

A barometer is a scientific instrument used to measure atmospheric pressure.

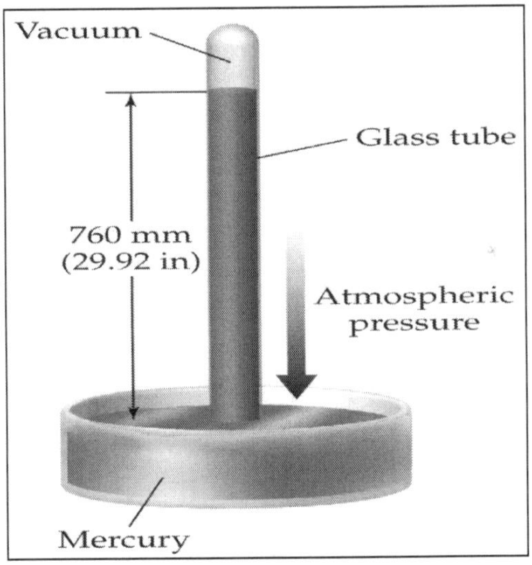

Figure 3.4. Simple Barometer.

a) Mercury Barometer

It is the oldest type of barometer, invented by the Italian physicist Evangelista Torricelli in 1643. When a one-meter long, open ended glass tube is filled with mercury and is turned upside down into a container filled with mercury, part of the mercury flows out of the glass tube into the container. "Torricellian vacuum" is then produced at the top of the glass tube and the mercury level stabilizes at approximately 76 cm from the mercury level in the container. Torricelli's experiment revealed that such a height indicates the ambient atmospheric pressure. The principle of mercury barometer is to measure atmospheric pressure from precise measurement of height. A mercury barometer has a vertical glass tube closed at the top sitting in an open mercury-filled basin at the bottom. Mercury in the tube adjusts until the weight of the mercury column balances the atmospheric force exerted on the reservoir. High atmospheric pressure places more force on the reservoir, forcing mercury higher in the column. Low pressure allows the mercury to drop to a lower level in the column by lowering the force placed on the reservoir.

b) *Fortin Barometer*

A portable mercury barometer that was invented in 1800 by the French instrument maker Jean Nicholas Fortin. In other mercury barometers air pressure is calculated from the distance between the level of the mercury in the reservoir and in the tube. In the Fortin barometer the bottom of the reservoir is flexible and can be raised or lowered by means of a screw, allowing the surface of the mercury to be adjusted to a predetermined level. A vernier scale on the side of the barometer tube is then lowered until its base touches the top of the mercury in the reservoir and a reading is taken from the fixed scale marked on the tube.

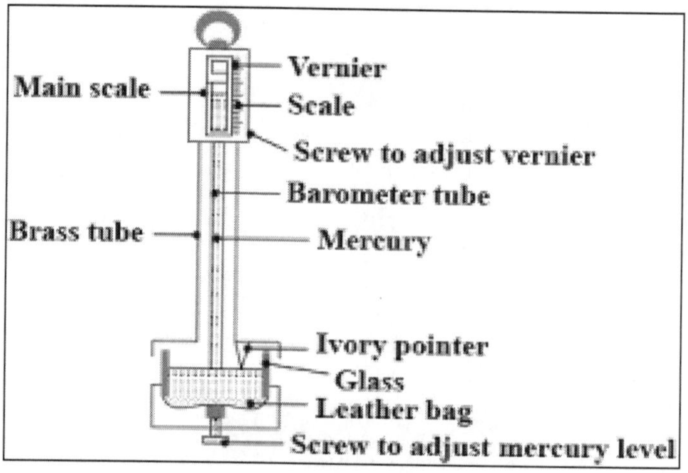

Figure 3.5. Fortin's Barometer.

Before travelling, the base of the reservoir is raised using the adjusting screw until both reservoir and tube are filled with mercury.

Any change in the atmospheric pressure is accompanied by an immediate change in the level of the mercury in the glass tube. As the height of the mercury column in the barometer changes, mercury flows between the tube and cistern. As a result, the level of the mercury in the cistern also changes. To determine the length of the mercury column in the barometer, it is necessary to know the position of the free surface in the cistern as well as in the tube. The first step in measuring atmospheric pressure using Fortin's barometer is to set the mercury level in the cistern. Using the adjustment screw, set the level of the mercury in the cistern such that the ivory pointer just touches the mercury. The reading of the top of the mercury column is then measured using both the main scale and the vernier scale. Before the readings are noted, the vernier scale needs to be positioned properly. The vernier scale is to be adjusted so that its edge and the corresponding reading in the main scale just set tangentially to the meniscus. Now, the readings on the main scale and the vernier scale are noted, and the atmospheric pressure is calculated.

Principle of Operation

The Fortin barometer allows the mercury level in the cistern to be set to a fixed datum or zero every time a reading is taken. The mercury is contained within a flexible pouch and a screw jack causes the level to either rise or fall.

c) Aneroid Barometer

An aneroid barometer is an instrument for measuring pressure as a method that does not involve liquid. Invented in 1844 by French scientist Lucien Vidi. Inside this instrument is a small, flexible metal box called an aneroid cell. Before the cell is tightly sealed, air is partially removed, so that small changes in external air pressure cause the cell to expand or contract.

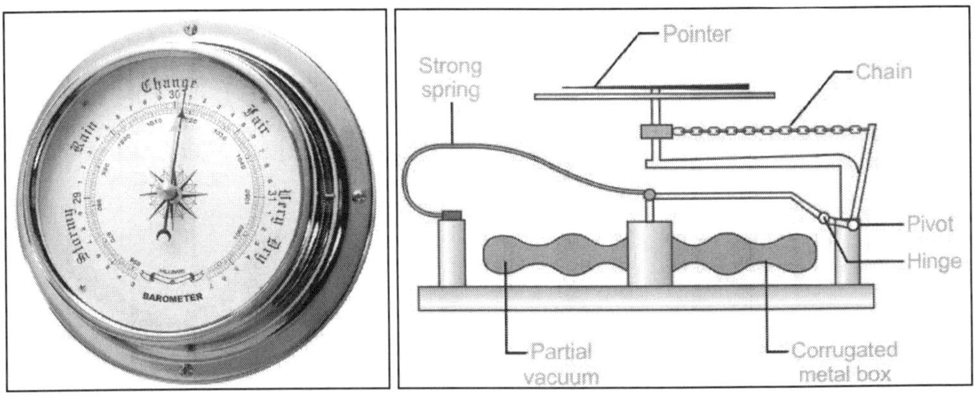

Figure 3.6. Aneroid Barometer.

The size of the cell is calibrated to represent different pressures, and any change in its size is amplified by levers and transmitted to an indicating arm, which points to the current atmospheric pressure the aneroid barometer often has descriptive weather-related words printed above specific pressure values. These descriptions indicate the most likely weather conditions when the needle is pointing to that particular pressure reading. Generally, the higher the reading, the more likely clear weather will occur, and the lower the reading, the better are the chances for inclement weather. This situation occurs because surface high-pressure areas are associated with sinking air and normally fair weather, whereas surface low-pressure areas are associated with rising air and usually cloudy, wet weather.

A steady rise in atmospheric pressure (a rising barometer) usually indicates clearing weather or fair weather, whereas a steady drop in atmospheric pressure (a falling barometer) often signals the approach of a cyclonic storm with inclement weather. The altimeter and barograph are two types of aneroid barometers. Altimeters are aneroid barometers that measure pressure, but are calibrated to indicate altitude. Barographs are recording aneroid barometers. Basically, the barograph consists of a pen attached to an indicating arm that marks a continuous record of pressure on chart paper. The chart paper is attached to a drum rotated slowly by an internal mechanical clock.

Principle

A metal box partially exhausted of air (vacuum) will undergo changes of shape as the outside pressure varies. These small movements of the box are transmitted to a pointer displayed on the face of the aneroid barometer by a system of levers. This type of barometer is common in aircraft and in boats.

d) Barograph

Barograph is an instrument that makes a continuous pen and ink trace of the changing atmospheric pressure on a chart. The barograph chart is also called a barogram. Barograms are weekly rectangular charts recording barometric pressure at a given station. A barograph is records the barometric pressure over time in graphical form.

Figure 3.7. Barograph.

This instrument is also used to make a continuous recording of atmospheric pressure. The pressure-sensitive element, a partially evacuated metal cylinder, is linked to a pen arm in such a way that the vertical displacement of the pen is proportional to the changes in the atmospheric pressure. The principle of the barograph is same as that of the aneroid barometer. A metal box partially exhausted of air will undergo changes of shape as the outside pressure varies. These small movement of the box are transmitted by a system of a levers to a recording arm that has at its extreme end of a pen. A pen records on paper using ink held in a nib. The recording materials is mounted on a cylindrical drum which is rotated slowly by a clock, commonly, drum makes one revolution per day, per week, per month or per week as per user use.

The principle of the barograph is same as that of the aneroid barometer. A metal box partially exhausted of air (vacuum) will undergo changes of shape as the outside pressure varies. These small movements of the box are transmitted by a system of levers to a recording arm that has at its extreme end either a scribe or a pen. A scribe records on smoked foil while a pen records on paper using ink, held in a nib. The recording material is mounted on a cylindrical drum which is rotated slowly by a clock. Commonly, the drum makes one revolution per day, per week, or per month and the rotation rate can often be selected by the user.

Chapter 4

Wind

4.1 Introduction

Wind is the horizontal flow of air and the vertical movement of air is called as air current. Windward side is the direction from which the wind comes and the direction towards the wind blows is called as leeward side. The wind velocity depends upon pressure gradient. The air moves horizontally from high pressure area to low pressure area along the pressure gradient. The direction of wind first is decided by the direction of pressure gradient.

4.2 Variation of Wind Speed

Unit of measurement of wind velocity is knot, m/s, km/hr, mile/hr and feet/sec.

1 knot = 1.15 miles/hr

= 0.5148 m/s

= 1.853 km/hr

= 1.687 ft/s

4.2.1 Daily Variation of Wind Speed

The wind has maximum speed in the early afternoon and minimum in the early morning before sunrise.

4.2.2 Seasonal Variation of Wind Speed

The wind speed maximum in South-West and North-East monsoon season and minimum in summer and winter season.

4.3 Forces Acting to Produce Wind

a) Pressure Gradient Force

It is the rate of change in atmospheric pressure between two points at the same elevation. It is also called as isobaric slope. This force causes the air movement in the direction of low pressure. It acts in the perpendicular direction of isobars. The change of pressure per unit horizontal distance is called pressure gradient force. The direction of flow of air is from high to low pressure and speed of flow is directly related to the pressure gradient.

b) Coriolis Force

This force results due to earth rotation around its axis. It causes the wind to deflect to its right direction in northern hemisphere and deflects to its left direction in southern hemisphere. Coriolis force is minimum at equator and maximum at poles. This force is directly proportional to the angle of latitude. As latitude increases and speed of the earth's rotation decreases, Coriolis force increases. The Coriolis effect changes wind direction but does not change wind speed. At equator, wind blows perpendicular to the isobars. The winds directly get uplifted vertically to form thunderstorm.

c) Centrifugal Force

The force comes into play, when air particles move in curved path, an acceleration towards the center of rotation. It makes to throw the particle away from center. The centrifugal force works against the gravitational attraction and directed forward the earth center.

d) Frictional Force

This force results due to friction between moving air and surface and it increases with surface roughness. It acts in opposite direction of wind motion and reduces the wind velocity. This is strongest near surface and become weak at higher altitude. At one km altitude, effect is negligible hence, wind speed increases with altitude.

4.4 Classification of Wind

A) Vertical Wind

a. Divergence
b. Convergence
c. Eddies
d Convection

B) Horizontal Wind

a. Periodic winds

b. Regular planetary earth circulation system winds
c. Local wind

a) Periodic Winds

Monsoon is a periodic circulation. The word monsoon is used for such a circulation, which reverses its direction after every six months *i.e.* from summer to winter and from winter to summer. In summer, the moist wind blows from ocean to land and in winter, cold and dry wind blows from land to ocean. Monsoon circulation involves 180° change in the direction of winds. Indian monsoon season consist of South-West monsoon wind and North-East monsoon wind.

b) Regular planetary earth circulation system winds

There are three types of winds, *i.e.* **trade winds**, **anti-trade winds** and **polar winds** which blow between different pressure belts over the globe.

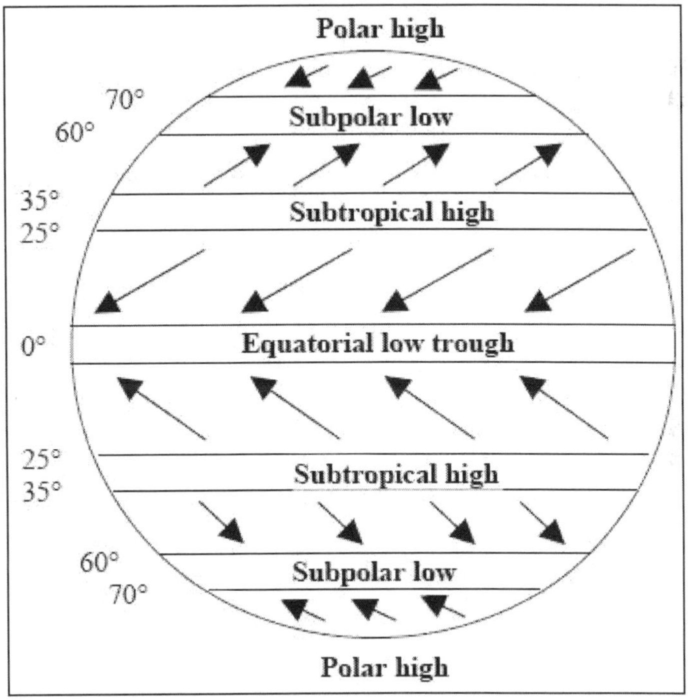

Figure 4.1. Regular Planetary Earth Circulation System Winds.

1. *Equatorial Low-Pressure Belt*

This low-pressure belt exists over equatorial region between 5°N and 5°S. It is known as belt of rising air or belt of convergence or doldrums (The belt with calm and variable winds). The average pressure in this belt is less than 1013 mbar but in eastern hemisphere it is less than 1009 millibar. This belt is thermally produced due to high insolation over the equatorial region. Intertropical convergent zone, it is zone at which the northeast and southeast winds meet and converge (5°N and 5°S).

Trade Winds blow between equational low pressure and subtropical high-pressure belts *i.e.* between 5° and 30° latitudes in both the hemispheres. Trade winds in Northern hemisphere are called as North-East trade winds and in Southern hemisphere are called as South-East trade winds.

2. *Sub-Tropical High-Pressure Belt*

This pressure belt lies between 25° and 35° N and S in both hemispheres. This belt is characterized by sinking of air from higher altitudes. Sinking of equatorial air in this zone might be because of two reasons. First cooling of equatorial air at higher level, results in increased density which cause sinking. Second due to earth rotation from west to east, pole ward directed winds are deflected east wards. It causes blocking effect and air piles up aloft and this causes a general sinking. It is also known as belt of divergence or horse latitudes. The sinking air currents create stable atmosphere with dry air masses. All the hot deserts of the world lie in this belt. The winds flow from the horse latitude to the equatorial region known as trade winds. While moving, these winds are deflected by Coriolis force to the right in Northern hemisphere and to the left in Southern hemisphere.

3. *Sub Polar Low-Pressure Belt*

This belt of low pressure lies between 60° and 70° latitude in both the hemispheres. It is also the belts of rising air/converging air as winds coming from the subtropics and polar areas converge and rise in this belt. This belt results due to rotational movement of earth, which swings up air in this zone. Due to a great contrast between the temperature of winds from subtropical and polar regions, cyclonic storms are developed in this region. The winds that flow from sub-tropical high to the sub polar low-pressure area are known as prevailing westerlies or antitrade winds. At 40°S to 65°S the winds are violent and have high frequency creating large noise hence this latitude is known as roaring forties.

Anti Trade winds blow between 35° and 60° latitudes in both hemispheres or they blow from subtropical high-pressure belt to sub polar low-pressure belt in each hemisphere. The direction of anti-trade winds in Northern hemisphere is North-West and South-West in Southern hemisphere. The westerlies are the zone of cyclonic storms in between 40° and 60° latitudes.

4. *Polar High-Pressure Belt*

This belt lies over the poles in both hemispheres. The high pressure in these belts is mainly due to extreme low temperatures persist in polar region which result in more dense and heavy air mass. The easterly winds blow out of these high-pressure belts towards sub polar low-pressure belt is known as polar winds. Polar winds blow between polar high pressure and sub polar low-pressure belts *i.e.* between 90° and 70° latitudes in both the hemispheres.

c) Local Wind

These winds originate due to differences in local features. Local wind is a tertiary circulation over a meso and microscale.

1. *Land Breeze*

At night time, a low pressure is developed over water surface and high pressure is developed over land surface. Due to this pressure gradient between land and sea, air moves from land to sea.

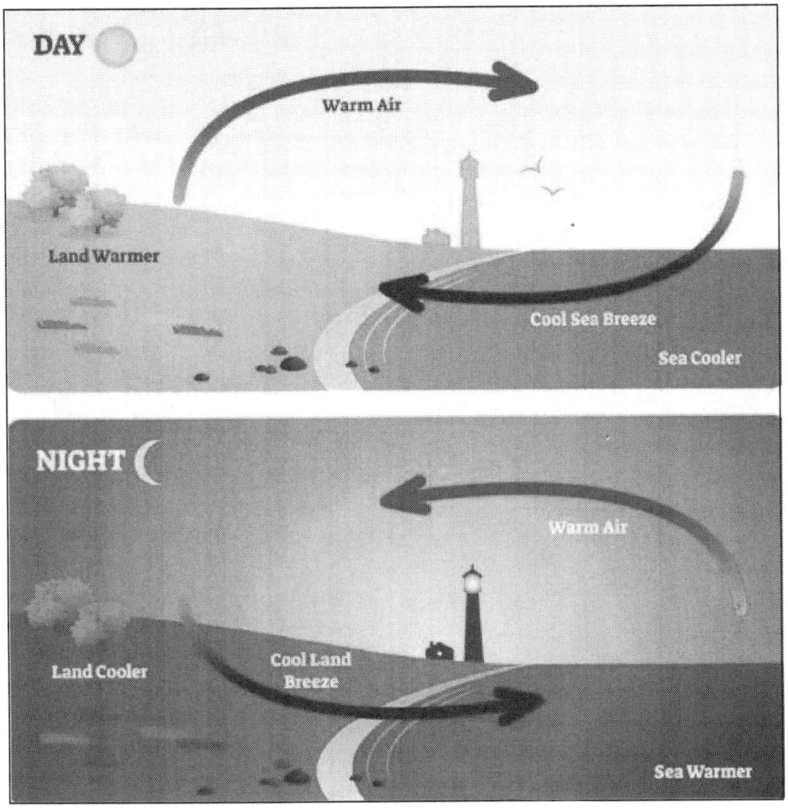

Figure 4.2. Land and Sea Breeze.

2. *Sea Breeze*

At daytime, a low pressure is developed over land surface and high pressure is developed over water surface. Due to this pressure gradient between land and sea, air moves from sea to land.

3. *Mountain Winds*

At night, the mountain tops radiate energy and cool to lower temperatures as compared to valley. The cool denser air moves down along the mountain slope into the valley. This air movement from mountain to valley is called mountain winds. It is also known as katabatic wind.

4. Valley Winds

At day time, the mountain top gets energy earlier and warms to temperatures higher than valley temperatures. The warm air over mountain top rises above and generates upslope flow of air from valley. This movement of air from valley to mountain top is known as valley wind. It is also named as anabatic wind.

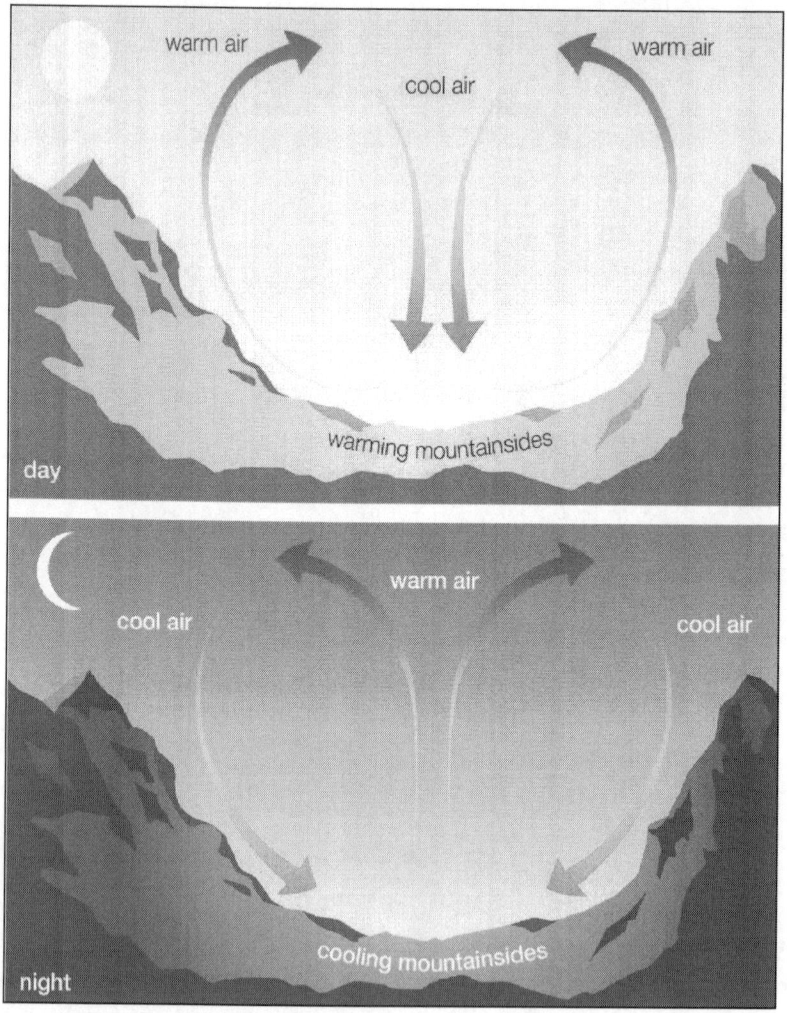

Figure 4.3. Mountain Valley Winds.

5. Warm Local Winds

These are hot and dry winds resulted due to advection of hot and dry air from warm places. *e.g.* Foehn is warm and dry wind on the northern side of Alps

Mountain range in Switzerland. Chinook is warm and dry wind on the eastern side of Rocky Mountains in North America. Santa Ana is warm and dry wind experienced in California in U.S.A. Sirocco is warm, dry and dusty wind blowing from Sahara Desert in Southern Italy. Khamsin is hot and dry wind experienced in Egypt. Simoon is a very hot and dry wind experienced in Asiatic and African deserts. Loo is a very hot and dry wind experienced in North-West India during summer. Harmattan is a hot, dry and dusty wind experienced in west coast of Africa.

6. *Cold Local Winds*

These are cold and dust laden winds occur in different conditions. *e.g.* Mistral is cold winds experienced in southern coast of France and Mediterranean coast. Bora is experienced in eastern shore of the Adriatic Sea. These winds also experienced in northern Black Sea Coast in Japan and Northern Scandinavia. Blizzard is strong winds with snow storms in great plains to the east of Rocky Mountains.

4.5 Circulation Cells

a) Hadley Cell

It is also termed as tropical cell. At low altitude air movement is towards the equator, where it gets heat and rises vertically with poleward movement at higher altitude. It is circulation of air between subtropical high-pressure belt and equatorial low-pressure belt in both the hemisphere. The surface winds in this cell are trade winds.

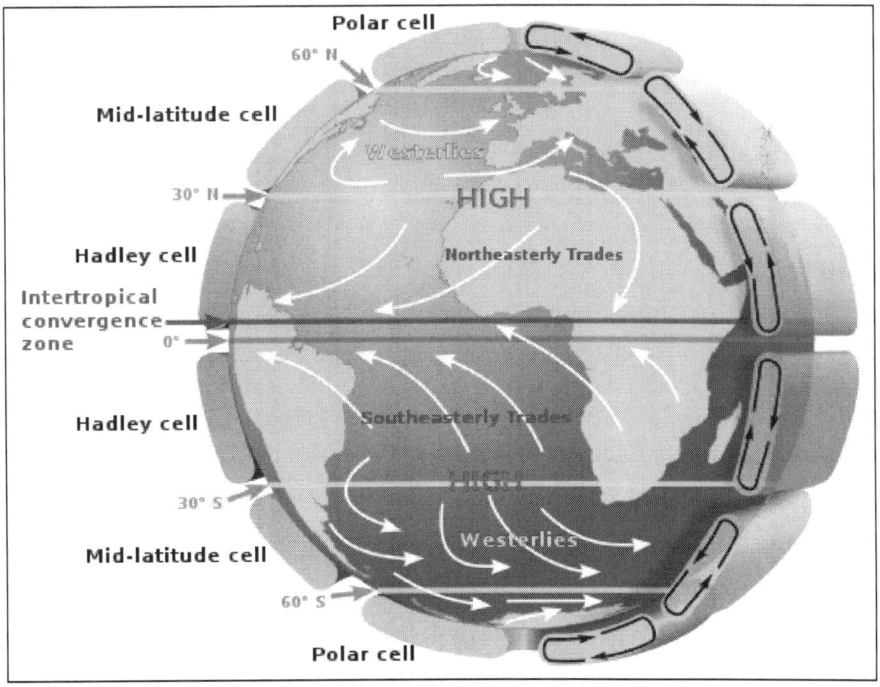

Figure 4.4. Circulation of the Wind.

b) Ferrel Cell

It is a mid-latitude mean atmospheric circulation cell between sub-tropical high-pressure belt and sub polar low-pressure belts. In this cell the air moves poleward and East ward near earth surface and equator ward and westward at higher altitudes. The surface wind in this cell is westerlies.

c) Polar Cell

In this cell air rises and moves poleward at higher levels and sinks at poles and diverges outward from the poles. This cell exists between polar high pressure and sub polar low-pressure belts. The surface winds in this cell is easterlies.

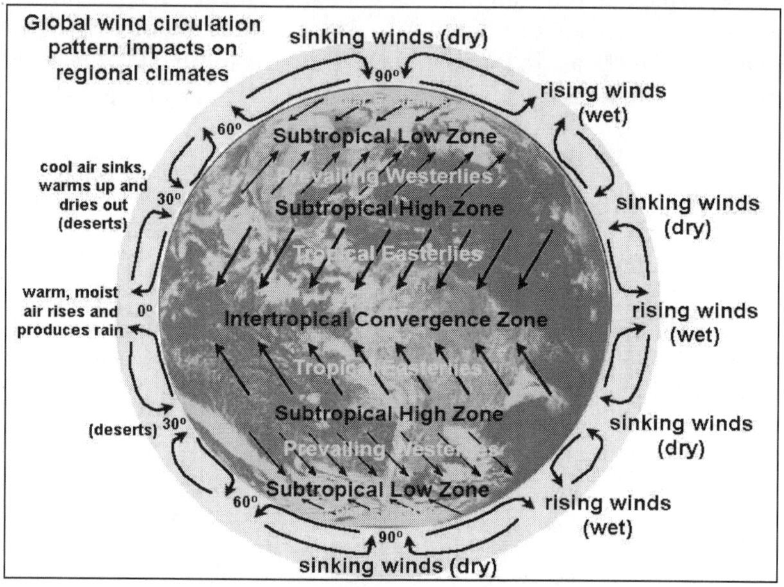

Figure 4.5. Circulation of the Wind.

d) Walker Cell

The vertical circulation in the tropical atmosphere is known as walker cell. It develops due to difference in sea surface temperature between east and west equatorial Pacific Ocean. Ascending branch of walker cell located over Indonesia and descending limb is located over the semi-arid regions of north-west India, Pakistan and middle of East.

4.6 Important Facts about Wind

- ☆ **Isotach** is the line on weather map, which joins the places of equal wind speed.
- ☆ **Vorticity** is a whirling motion of fluid particles.
- ☆ **Vortex** is a whirling mass of fluid or air, especially a whirlpool or whirlwind.

- **Haze** is a mixture of dust and smoke particles.
- **Dust devil** is a whirling dust column of height 1 km or more over the earth surface.
- **Gale** is a wind having velocity between the ranges of 28-55 km.
- **Squall** is sudden increase in wind speed (>50 km/hr or Beaufort scale up to 6 or more) for at least 1 min and then decreases.
- Convection is the vertical rise of warm and light air near the ground which results due to heating of ground.
- **Gust** is an eddy moving down with mean speed greater than 25 to 30 km/hr. It is an increase in the strength of wind for a short time due to increase of turbulence.
- **Lull** is an eddy moving up with mean wind speed less than 5 km/hr.
- **Storm surge** is the unusual rise in the water level generated by storm over and above the predicted astronomical tide.
- **Seiche** is a wave on the surface of a lake or landlocked bay caused by atmospheric or seismic disturbances.
- **Geostropic wind** is the wind, which blows parallel to isobars in a straight line, with a balance between Coriolis force and pressure gradient force acting on it. or when the wind flows in a **straight line** with no frictional force on it, the only force acting is the Coriolis and pressure gradient forces. Its velocity is inversely proportional to the distance between the isobars.
- **Gradient wind** is the wind that results with a balance between pressure gradient force on one hand and the Coriolis force and centrifugal force on the other hand. It moves parallel to isobars but in **curved path** while geostrophic wind moves parallel to straight isobars.
- **Thermal wind** is the vector difference between the winds at the top and bottom of any layer.
- **Windward** is the direction from which the wind blows.
- **Leeward** is the direction to which wind flows.
- When the direction of wind changes continuously in clockwise direction is called **veer** or **haul** and when the direction of wind changes continuously in anticlockwise direction is called **back.**
- **Eddy** is a whirl or circular spinning small current of air or water that is different from the general flow of air.
- **Ekman spiral** is a structure of currents or winds near a horizontal boundary in which the flow direction rotates as one moves away from the boundary. An Ekman spiral is a rotating column of water that forms when water moves at an angle to the wind direction due to the Coriolis effect.

- **Tornado** is a most violent and destructive whirlwind. It is funnel shaped and its diameter is very small *i.e.* near about 10 m and wind velocity is about 460 mile/hr.
- A **tornado** is a rapidly rotating column of air that is in contact with both the surface of the Earth and a cumulonimbus cloud.
- **Jet streams** is geostrophic wind blowing horizontally through the upper layer of the troposphere generally blow from **west to east**. It is fast blowing air current system in upper atmosphere (9-18km), its speed is maximum 340 km/hr in the middle parts. It is extended from 20° latitude to the poles in both hemispheres.
- The whirl movement of the Jet stream is called **Rossby wave**.
- **Buys Ballots law:** In northern hemisphere, if we stand with our face towards the wind, the pressure will be lower on the right hand and in southern hemisphere, lower will be on the left hand.
- Wind speed **more than 50 km/hr** leads to lodging of crop leading to heavy loss.
- **Ferrel law**: The earth rotates from west to east and due to rotation, air particles moving in a straight line in a northern hemisphere is deflected towards its right and movement becomes clockwise, In the southern hemisphere the deflection is towards left and winds move anticlockwise.
- Wind stronger than **8 km/hr** can loft sand in arid regions leading to formation of sand dunes.
- The addition of salt in the soil in the coastal areas through wind is known as **pulverization**.
- Wind turning to right form its course (clockwise direction) is called **veering wind**.
- Wind turning to light form its course (anti-clockwise direction) is called **backing wind**.
- **Wind shear** is the variation of any of the components of wind at a given location.

4.7 Instruments Used to Measure Wind Speed and Direction

a) Wind Vane

The Greek astronomer Andronicus built the first wind vane for the tower of winds in Athens. It is made of bronze. It is an instrument used for showing the direction of the wind. It is balanced lever. One end of the lever exposes a broad surface consisting of a flat plate (fang). The other end is narrow and point to the direction from which the wind is blowing (pointer). The entire rotating assembly is firmly attached to the bearing housing cap. Below the bearing housing cap is the top ball bearing housed within the support tube. The spindle is brass rod and horizontal arm moves the spindle. Below this moving system, a grid cross arm is fixed and set to four directions and divided into 8 or 16 directions with the help of pointed rods. It is fixed over the wooden post at the height of 3 meters from the ground in open terrain or on the top of the building.

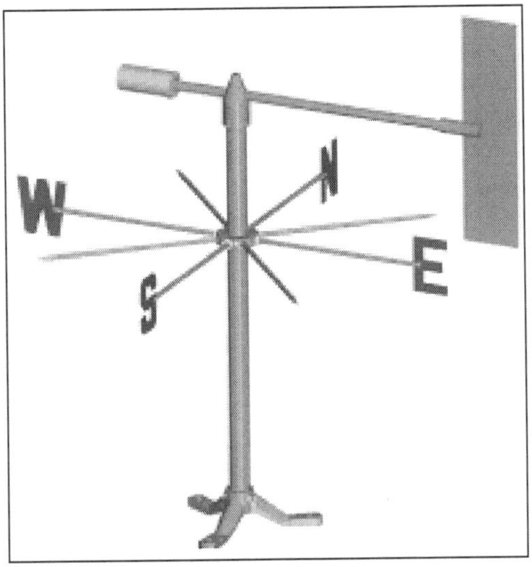

Figure 4.6. Wind Vane.

b) Robinson Cup Anemometer

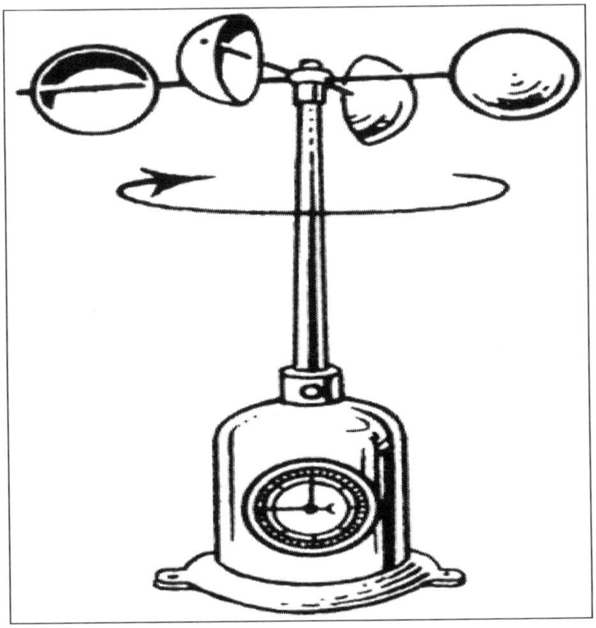

Figure 4.7. Robinson Cup Anemometer.

Anemometer is the instrument used for measuring wind speed. The term derived from the Greek word *anemos* which means wind. It is a simple type of anemometer was invented in 1845 by John Thomas Romney Robinson. It consisted of four hemispherical cups mounted on horizontal arms, which were mounted on a vertical shaft. The air flow past the cups in any horizontal direction turned the shaft at a rate that was roughly proportional to the wind speed. Therefore, counting the turns of the shaft over a set time interval produced a value proportional to the average wind speed for a wide range of speeds. It is also called a rotational anemometer. It is not sensitive to record sudden changes in wind velocity such as squalls, gusts *etc.* The three cup anemometer developed by the Canadian John Patterson in 1926.

Figure 4.8. Cup Anemometer.

The three-cup anemometer was further modified by the Australian Dr. Derek Weston in 1991 to measure both wind direction and wind speed. It is installed on a metal pipe, which is fixed on a wooden post. The height from the centre of the anemometer cups is 10 ft above the ground level.

Conversion Units of Wind Speed

Units	*Knot*	*m/s*	*Kmph*	*mph*	*Ft/s*
Knot	1	0.514	1.852	1.151	1.688
m/s	1.944	1	3.6	2.237	3.281
Kmph	0.540	0.278	1	0.621	0.912
mph	0.869	0.447	1.609	1	1.466
Ft/s	0.592	0.305	1.097	0.681	1

Unit of Measurement

Meters second^{-1}, kilometers hour^{-1}, miles per hour, feet per second and knots

Observation

The reading of wind speed is obtained from the cyclometer dial having a range of 0–9999. The four black figures give whole km and the red figure to the right gives tenth of km. Wind speed is calculated as follows;

- ☆ Note down two reading from anemometer at an interval of 3 min. Multiply the difference by 20 to get wind speed at the time of observation in kmph.
- ☆ Subtract the anemometer reading at 07.00 h LMT of the previous day from that at 07.00 h LMT of the observation day and divide the difference by 24 to get the mean daily wind speed for the observation day in kmph.

Examples

1. Calculate the wind speed in knots, if the first anemometer reading is 4005.6 kmph and after 3 minutes the reading is 4008.8 kmph.

   Wind speed (kmph) = 20 (b – a)
   = 20 (4008.8 – 4005.6)
   = 20 x 3.2
   = 64 kmph

   Wind speed = Wind speed (kmph) x 0.54
   = 64 x 0.54
   = 34.56 knots

2. Calculate the mean daily wind speed in knots, if anemometer reading on 14.12.2022 at 0700 hrs is 3345.9 kmph and reading on 15.12.2022 at 0700 hrs is 3675.5 kmph.

   Wind speed (kmph) = (b – a)/24
   = (3675.5 – 3345.9)/24
   = 329.6/24
   = 13.73 kmph

   Wind speed = Wind speed in knots x 0.54
   = 13.73 x 0.54
   = 7.41 knots

c) Propeller Anemometer

A propeller anemometer has a sensor with a streamlined body and a vertical tail to detect wind direction and a sensor in the form of a propeller to measure wind speed integrated into a single structure. It measures wind direction and wind speed, and record the instantaneous wind direction and wind speed in remote locations. It also measures the average wind speed using wind-passage contacts or by calculating the number of optical pulses.

Figure 4.8. Propeller Anemometer.

d) Wind Sock

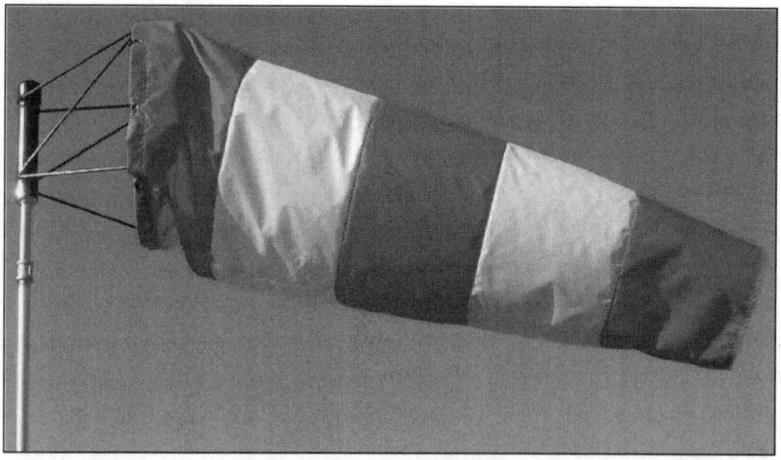

Figure 4.8 Wind Sock

A windsock, also known as a wind cone, is a conical tube designed to indicate wind direction and relative wind speed. They are made from light, resilient materials such as nylon or PVC and are often brightly colored or have reflective surfaces to enhance visibility. They are open at one end and closed at the other. The open end is attached to a pole or mast, while the closed end is left free to move with the wind. When the wind blows, it inflates the windsock, causing it to fill with air and extend in the direction from which the wind is coming. The direction in which the windsock is pointing indicates the wind's direction. They can also provide to estimate wind speed based on the degree of inflation and movement. Windsock usually has five stripes (red/white or orange/green) and each stripe represents 3 knots of wind *i.e.* when the first stripe extends, the wind speed reaches 3 knots. When the second stripe extends, it is 6 knots, and so on, until all stripes are extended when the wind speed is 15 knots or more.

e) Digital Anemometer

It is a device that measures wind speed and direction.

Figure 4.9. Digital Anemometer.

f) Wind Rose

It is the pictorial representation of wind direction and wind speed. It has 16 directions with 22.5° differences between each direction. Presented in a circular format, the length of each "spoke" around the circle indicates the amount of time that the wind blows from a particular direction. Colors along the spokes indicate categories of wind speed.

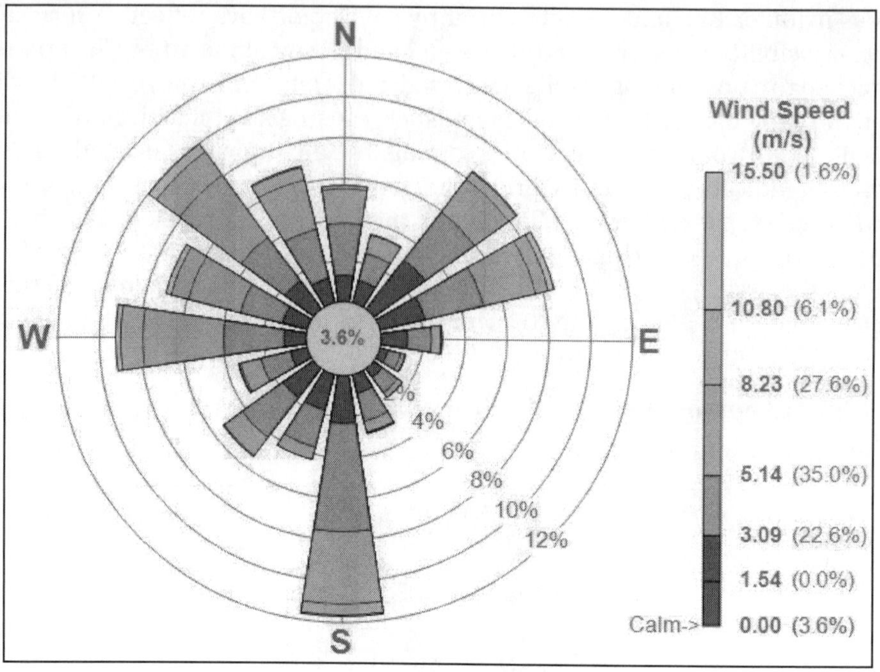

Figure 4.10. Wind Rose.

Beaufort's Scale

It estimates the velocity and strength of the wind. This wind scale was designed by Admiral Beaufort in 1805. The approximation formula for conversion of Beaufort scale number into meters per second is;

$$2 \times \text{Beaufort No.} - 1 = \text{Velocity } (ms^{-1})$$

Beaufort No.	*Types of Wind*	*Effect Produced on the Objects*	*Speed of Wind (miles/hr)*
0	Calm	Smoke rises vertically	Less than 1
1	Light air	Smoke drifts, wind vane does not move	1-3
2	Light breeze	Wind felt on face, leaves rustler, wind vane moves	4-7
3	Gentle breeze	Leaves and twigs in motion, wind extends light flag	8-12
4	Moderate breeze	Raise dust and loose papers, small branches are moved	13-18
5	Fresh breeze	Small leaves in trees way, wavelets on pond water	19-24

Beaufort No.	*Types of Wind*	*Effect Produced on the Objects*	*Speed of Wind (miles/hr)*
6	Strong breeze	Large branches in motion, whistling heard in telegraph wires, umbrellas used with difficulty	25-31
7	Moderate gale	Whole tree in motion, inconvenience in walking against wind	32-38
8	Fresh gale	Breaks twing from trees	39-46
9	Strong gale	Slight structural damages	47-54
10	Whole gale	Trees, uprooted, considerable structural damages	55-63
11	Storm	Very rarely experienced, wide spread damage	64-75
12	Hurricane	Violence and destruction	75-136

Chapter 5

Temperature

5.1 Introduction

Temperature is the degree of hotness and coldness. The interaction of insolation with the atmosphere and the earth surface creates heat which is measured in terms of temperature. When any object receives energy in the form of heat it's temperature increases. Heat is the energy which makes things or objects hot, while temperature measures the intensity of heat. Air or atmosphere receives the heat energy from the sun and its temperature increase.

Minimum and maximum temperatures that define limits of growth and development of a plants, and an optimum temperature at which growth proceeds with greatest rapidity is known as cardinal temperature. Minimum temperature is the temperature below which no growth occurs. For typical cool season crops, it ranges between 0-5°C and for hot season crops; it is between 15-18°C. Optimum temperature is the temperature at which maximum plant growth occurs. For cool season crops, it ranges between 25-31°C and for hot season crops between 31-37°C. Maximum temperature is the temperature above which the plant growth stops. For cool season crops, it ranges between 31-37°C and for hot season crops; it ranges between 44-50°C. Low temperature affects several aspects of crop growth, *viz.*, survival, cell division, photosynthesis, water transport, growth and finally yield and high temperature adversely affects mineral nutrition, shoot growth and pollen development resulting in low yield.

5.1.1 Heat Injuries

Thermal death point is the temperature at which the plant cell gets killed when the temperature ranges from 50-60°C. High temperature results in desiccation of plants, disturbs the physiological activities like photosynthesis and respiration and increases respiration leading to rapid depletion of reserve food.

a. Sun Clad

It caused injury on the barks of stem by high temperature during day time and low temperature during the night time.

b. Stem Griddle

The stem at ground level scorches around due to high soil temperature. It causes death of plant by destroying conductive tissues. *e.g.* This type of injury is very common in young seedlings of cotton in sandy soil when soil temperature exceeds 60°C.

c. Scorching

High temperature leads to dehydration and leaves are scorched.

5.1.2 Cold Injury

a. Chilling Injury

Plants which are adapted to hot climate, if exposed to low temperature are found to be killed or severely injured or development of chlorotic condition (yellowing). *e.g.* chlorotic bands on the leaves of sugarcane, sorghum and maize in winter months when the night temperature is below 20°C.

b. Freezing Injury

When the plants are exposed to very low temperature, water freezes into ice crystals in the intercellular spaces of plants. The protoplasm of cell is dehydrated resulting in the death of cells.

c. Suffocation

In temperature regions, during the winter season, the ice or snow forms a thick cover on the soil surface. As a result, the entry of oxygen is prevented and crop suffers for want of oxygen. Ice coming in contact with the root prevents the diffusion of CO_2 outside the root zone. This prevents the respiratory activities of roots leading to accumulation of harmful substances.

d. Heaving

This is a type of injury caused by lifting up of the plants along with soil from its normal position. This type of injury is commonly seen in temperate regions. The presence of ice crystals increases the volume of soil. This causes mechanical lifting of the soil.

5.2 Measurement of Temperature

To measure the temperature Galio invented the thermometer in 1590. Also, there is number of types of thermometers such as thermometer in Stevenson screen, sling thermometer, Assmans Psychrometer and thermograph. There are four scales of temperature as given below;

Scale	*Invented by*	*Invented in Year*	*Boiling Point*	*Freezing Point*
Celsius (°C)	Anders Celsius	1742	100°C	0°C
Kelvin (°K)	Baron Kelvin	1848	373°K	273°K
Fahrenheit (°F)	D.G. Fahrenheit	1714	212°F	32°F
Reaumer (°R)	R.A. Reaumur	1730	80°R	0°R

5.3 Conversions of Temperature

- $°C = °K - 273$
- $°C = \frac{5}{9} \times (°F-32)$
- $°K = \frac{5}{9} (°F-32) + 273$

Example

a) Convert 26 °C (Celsius) to °F (Fahrenheit)

$°F = (°C \times 9/5) + 32$

$°F = (26 \times 9/5) + 32$

$°F = (46.8) + 32$

$°F = 78.8°F$

b) Convert 98°F (Fahrenheit) to °K (Kelvin)

$°K = (°F - 32) \times 5/9 + 273$

$°K = (98 - 32) \times 5/9 + 273$

$°K = 66 \times 5/9 + 273$

$°K = 309.67°K$

5.4 Air Temperature

The average global temperature is 15°C. Maximum temperature is the highest temperature of the day and recorded at 1400 hrs. Minimum temperature is the lowest temperature of the day and recorded at 0700 hrs. Temperature lag or thermal inertia is the amount of time between incoming energy maxima and temperature maxima.

5.4.1 Variation in Air temperature

5.4.1.1 Daily Temperature Variation

The maximum air temperature is received at the noon and recorded between 1400 hrs and 1500 hrs. This delay in time to reach maximum temperature is known as thermal lag. The minimum temperature occurs just before sunrise and recorded at 0700 hr.

5.4.1.2 Seasonal Temperature Variation

The highest temperature recorded in summer season and the lowest temperature recorded in winter season. The annual temperature range varies from place to place. There is thermal lag of 30-40 days. Solar radiation is maximum in July-August months and minimum in January-February months. The difference between the highest and lowest temperature for a given period is known as temperature range.

5.4.1.3 Horizontal Temperature Variation

It is the distribution of temperature across the latitudes over the surface of the earth. The rate of change of temperature with a horizontal distance is known as temperature gradient. The horizontal distribution of temperature is shown by isotherm. Isotherms are line connecting points that have an equal temperature. The maximum solar energy is received in equatorial region hence, the highest temperature is observed in equatorial region. As the latitude increases the solar energy received on earth surface decreases. Temperature decreases with increase in latitude being lowest on the pole. The sun crosses the equator twice in year hence, two maxima and two minima are observed in annual cycle. Outside this zone only one maxima and one minima are observed. The horizontal distribution of temperature over the globe studied from the isotherm of January and July months in both northern and southern hemispheres. In January, the sun shines vertically overhead near the tropic of Capricorn. Hence, it is summer in southern hemisphere and winter in the northern hemisphere. In July, the sun shines vertically overhead near the Tropic of cancer. Hence, high temperatures are found in the entire northern hemisphere.

5.4.1.4 Vertical Temperature Variation

There is fairly regular decrease in temperature of the air with height above the earth surface. The average rate of decrease of temperature in troposphere is about 6.5°C/km is known as lapse rate.

Troposphere: 0 to 11 km, temperature decreases from 15 °C to -60°C.

Stratosphere: 11 to 50 km, Temperature increases from -60 °C to 0°C.

Mesosphere: 50 to 80 km, temperature decreases from 0°C to -90°C.

Thermosphere: above 80 km, temperature increases from -90°C to 950°C.

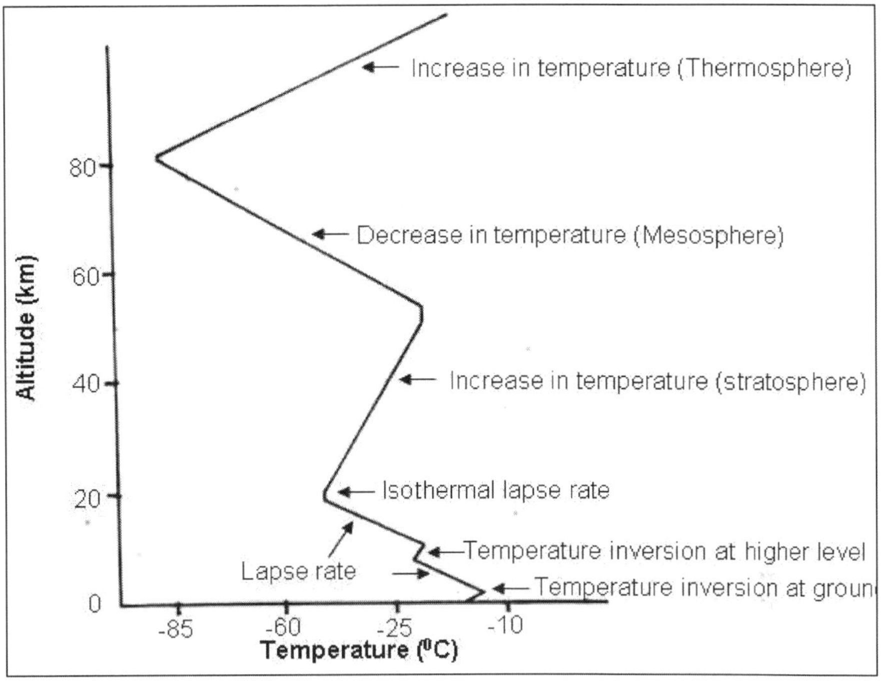

Figure 5.1. Vertical Temperature Variation.

5.4.2 Factors Affecting Temperature Distribution

- **Latitude:** Air temperature decreases with increase in latitude over the earth surface. Therefore, the places near equator are warmer as compared to the places in mid latitudes and near poles. Time of occurrence of maximum and minimum monthly mean temperature also depends on latitude.
- **Slope of the land:** Slope of the land controls the radiation intensity over a place, which in turn controls the temperature. Therefore, south facing slopes are warmer than north facing slopes in northern hemisphere and vice-versa in southern hemisphere.
- **Prevailing winds:** Winds can moderate the surface temperature of the continents and oceans. In the absence of winds, we feel warm in hot climates. At the same time, the weather is pleasant if wind blows.
- **Distribution of land and water:** Land surface is heated more quickly than water when subjected to equal amount of radiation and it also cools more rapidly. Variation in air temperature is much greater over the land than over the water. The differential heating process between land and sea surface are due to their properties. It is one of the causes for Indian monsoon.

- ☆ **Distance from the sea:** The distance from sea also affects the temperature distribution of a place. Diurnal range of temperatures is higher over places, which are far away from sea in comparison with costal places. The sea gets heated slowly and loses heat slowly where land het up and cool down quickly therefor the variation in temperature over the se is less compared to land.
- ☆ **Altitude:** The air temperature decreases with increase in altitude from the mean sea level as the density of air decreases and atmosphere is heated by terrestrial radiation therefore the places near the sea level record higher temperature than the places situated at higher altitude. This decrease in temperature with altitude ceases at tropopause. The normal lapse rate is 6.5°C/km.
- ☆ **Nature of soil:** Rocky and sandy soils warm more quickly and cool more rapidly as compared to heavy soils, which contain more water.
- ☆ **Ocean currents:** The energy received over the ocean surface carried away by the ocean currents from the warm areas to cool areas. This results in temperature contrast between the equator and poles. The occurrence of El-Nino is due to change in sea surface temperature between two oceanic regions over the globe. The places located on the coast where the warm ocean current flow record higher temperature than the places located on the coast where the cold currents flow.
- ☆ **Vegetation:** The vegetation cuts off much of the incoming solar radiation and also converts much of the energy into latent heat of evapo-transpiration. So, the places with more vegetation are cooler than the places without vegetation.
- ☆ **Mountain barriers:** Air at the top of the mountain makes little contact with the ground therefore it is cold and in the valley at the foothills makes a great deal of contact therefore it is warm. That is, the lower region of the earth's atmosphere is relatively warmer when compared to hillocks.
- ☆ **Condition of atmosphere:** Clouds absorbs the terrestrial radiation; therefore, cloudy nights are warmer than clear nights in winter. Dust particles also increase the temperature of atmosphere. Dusty atmosphere is warmer than clear atmosphere. A thick cloud reduces the amount of insolation received at a particular place and thus the day time temperature is low.

5.5 Adiabatic Change

It is any change in volume, pressure and temperature of air, when it moves upward or downward without any addition or subtraction of heat from the system.

a) **Adiabatic cooling:** It is decrease in temperature of air with its upward movement without subtraction of heat.

b) **Non adiabatic cooling:** It is decrease in temperature of air due to loss of heat by radiation, mixing of cold masses.

5.6 Lapse Rate

It is decrease in temperature with increasing height. The mean value of lapse rate in troposphere is 6.5°C/km or 3.5°F/1000 ft and in mesosphere is 5°C/km. The rate of which temperature decreases in a rising expanding air parcel called adiabatic lapse rate. No change in temperature with altitude known as Isothermal.

a. Dry Adiabatic Lapse Rate (DALR)

It is the rate of fall of temperature of unsaturated air parcel during adiabatic cooling. The DALR is about 9.8°C/Km or 5.39°F/1000 ft.

b. Saturated Adiabatic Lapse Rate (SALR)

It is the rate of fall of temperature of saturated air parcel during adiabatic cooling. The SALR is about 4°C/Km or 2.2°F/1000 ft.

5.7 Temperature Inversion

It is the increase in temperature with increase in height. Inversion is usually of short duration but quite common nonetheless. A long winter night with clear skies and still air is ideal situation for inversion. The heat of the day is radiated off during the night, and by early morning hours, the earth is cooler than the air above. Over polar areas, temperature inversion is normal throughout the year.

Surface inversion promotes stability in the lower layers of the atmosphere. Smoke and dust particles get collected beneath the inversion layer and spread horizontally to fill the lower strata of the atmosphere. Dense fogs in mornings are common occurrences especially during winter season. This inversion commonly lasts for few hours until the sun comes up and beings to warm the earth. The inversion takes place in hills and mountains due to air drainage. Cold air at the hills and mountains, produced during night, flows under the influence of gravity. Being heavy and dense, the cold air acts almost like water and moves down the slope to pile up deeply in pockets and valley bottoms with warm air above. This is called air drainage. It protects plants from frost damages.

Temperature inversion occurs in following conditions;

a) Earth surface loses heat more than it gains;

b) In mountain valley region and

c) Cold air masses replace warm air masses.

Types of Temperature Inversion

a) Surface Inversion

☆ **Radiation inversion** is due to radiational cooling of the ground. Air near the ground is cooler in winter while above its air is warmer.

- **Advection inversion** is a transport of energy in horizontal plane in downward direction occurs from an actual warming layer over a cold one.

b) Upper Air Inversion

- **Subsidence inversion** is occurring when a thick air mass moves downwards. It is common in high pressure area.
- **Convective inversion** is occurring above the surface by mechanical process. Due to frictional forces, eddies from the lower of atmosphere transport lower air to higher levels and bring back the upper air to lower level.

5.8 Stability and Instability of Atmosphere

a) **Stable atmosphere:** air parcel has tendency to return to its earlier position when the motivating force removed.

b) **Neutral atmosphere:** air parcel has tendency to moves to the new position when the motivating force removed.

c) **Unstable atmosphere:** air parcel has tendency to continue moving away from its original level when the motivating force removed.

DALR	*SALR*			*Condition of Atmosphere*
> 10°C	> 5°C	ELR > DALR	ELR > SALR	Unstable
= 10°C	= 5°C	ELR = DALR	ELR = SALR	Neutral
< 10°C	< 5°C	DALR > ELR	SALR > ELR	Stable

5.9 Instruments Used to Measure Air Temperature

Maximum Thermometer

It is a mercury thermometer used to measure the maximum temperature within a certain period. It has a narrow part in the capillary tube where mercury passage is constricted between the bulb and the starting point of the scale. As the air temperature rises, the mercury exits the bulb and passes through the constriction. When the air temperature falls, the mercury column breaks at this point. Thus, the mercury in the capillary tube cannot return to the bulb, and remains in the column indicating the maximum temperature. The maximum temperature is observed in the world between 1400 to 1600 hours of the day. The range of the thermometer is -35 to +55°C. After measurement, the thermometer is held at the head and the mercury in the capillary tube is shaken back into the bulb to reset its indication to the current air temperature. The reading of the maximum thermometer after setting should agree with that of dry bulb thermometer within 0.3°C.

Minimum Thermometer

It is an alcohol or spirit thermometer used to measure the minimum temperature within a certain period. It has a dumbbell-shaped index of colored glass in the spirit column. As the air temperature falls, the index is dragged by the surface tension of the spirit and moves toward the bulb with the top of the column. When the temperature rises, the index is left in position because the spirit flows through it. As a result, it remains in the column indicating the minimum temperature. After measurement, the column is inclined while keeping the bulb higher than the head, and the index is gradually slid back to the top of the column. The range of the thermometer is -40 to +50°C. The minimum temperature is observed at 0700 hrs of the day. After setting, the reading of the minimum thermometer should agree with that of dry bulb thermometer within 0.6°C.

Figure 5.2. Single Stevenson Screen.

Dry Bulb Thermometer

This is the mercury thermometer which is used to record dry air temperature. The temperature range is from -35°C to +55°C. It is called dry bulb temperature because air temperature is indicated by a thermometer not affected by the moisture of the air. This temperature used for calculating humidity, vapour pressure and dew point.

Wet Bulb Thermometer

The wet bulb temperature is the temperature of a parcel of air cooled to saturation by the evaporation of water into it, with the latent heat supplied by the parcel. It is mercury thermometer used to measure the moist air temperature. It is an ordinary thermometer with the bulb wrapped in wet muslin. The bulb acts as an evaporating surface. The bulb is kept continuously wet by providing water by means of four strands of cotton thread dipped into a small water container with distilled water or rain water. The adiabatic evaporation of water from the thermometer bulb results in the cooling effect and thus the wet bulb temperature is lower than the dry bulb temperature in the air. At 100 per cent relative humidity, the wet bulb temperature is equal to the dry bulb temperature; at lower humidity the wet bulb temperature is lower than dry bulb temperature because of evaporative cooling. The rate of evaporation from the muslin cloth on the bulb and the temperature difference between the dry bulb and wet bulb depends on the humidity of the air. The evaporation from the wet muslin is reduced when air contains more water vapor. The difference in temperature of dry and wet bulb known as wet bulb depression. This temperature is used to find out dew point temperature, vapor pressure and humidity.

Stevenson Screen

This Screen was designed by Thomas Stevenson in 1866. It is a specialized shelter to accommodate the four thermometers *i.e.*, maximum, minimum, dry bulb thermometers and wet bulb thermometers for recording of air temperature. Maximum and minimum thermometer are place in horizontal position on the upper and lower sides of wooden box, respectively and the bulb end rest at an angle of 2° to horizontal plane. Dry and wet bulb thermometers are kept vertical in the wooden box on the left and right sides, respectively. Stevenson's screen is a wooden rectangular box of the following dimensions; Single Stevenson's Screen (Length 56 cm, width 30 cm and height 40 cm) and Double Stevenson's Screen (Length 112 cm, width 30 cm and height 40 cm) with a double-layered roof and louvered sides. The screen is painted white and is mounted on four wooden supports. The support of the screen is placed at a height of 4 ft. (1.22 m) above the ground. The screen is set up with its door facing north side (opening downward) so that minimum sunlight would enter while the observer is reading the instruments. The screen is meant to protect the thermometers from direct heating from ground and neighboring objects and from losing heat by radiation at night. The double layered roof and louvered sides protects the instruments from rain and snow. It also allows free air circulation. The dry bulb and wet bulb thermometers are kept vertical on the wooden bracket on the left- and right-hand sides respectively.

Thermograph

A mechanical thermograph is an automatic temperature recording instrument having a rotating chart mechanism which provides a continuous plot of temperature against time.

A bimetallic thermograph is a unit consisting of a bimetal (bimetallic strip) and a clock-driven drum on which a recording chart is wound. The curvature of the bimetal changes in response to temperature variations, and this curvature change is recorded on a chart. The bimetal consists of two metal plates with different expansion coefficients welded or brazed and then rolled to an appropriate thickness. As the temperature changes, the bimetal curves due to the difference in the expansion coefficients of the two metals. This change is mechanically magnified and then recorded.

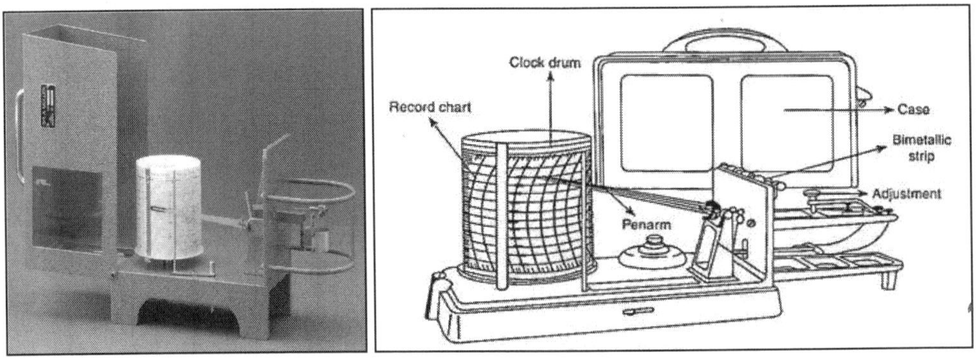

Figure 5.3. Thermograph.

When the bimetal curves in response to temperature change, the bimetal lever attached to the end of the bimetal moves with it. Changes in air temperature cause changes in the curvature of circular bimetallic strip creating micro motions. These micro motions are conveyed to the short arm of the lever system which amplifies these motions. The longer arm of the lever is provided with pen and ink and makes marking on the graph paper fixed on the rotating drum. The drum is rotated once in 24 hours or once in a week but the clock work inside. On the graph paper, temperature is on Y axis and time is on X axis.

Grass Minimum Thermometer

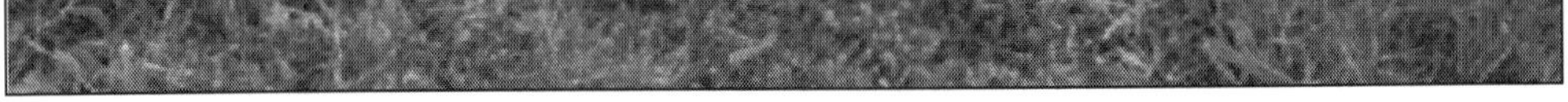

Figure 5.4. Grass Minimum Thermometer.

The grass minimum temperature is the lowest temperature reached overnight by a thermometer freely exposed to the sky just above short grass. The stem of this thermometer is however encased in a glass jacket in order to protect seal marks on the stem from dew, rain and prevent the stem from cooling rapidly. Grass minimum temperature indicates the chance of occurrence of ground frosts.

A ground frost is likely to occur when the instrument records a temperature below 0°C. Thermometers are mounted on suitable Y-shaped supports so that they are inclined at an angle of about 2° from the horizontal position, with the bulb being lower than the stem. The reading is taken before sunrise. After reading, the instrument is kept indoors to avoid direct exposure to solar radiation. The grass minimum temperature remains lower than the air temperature at screen level.

Assmann Psychrometer

It is used to determine the air temperature and humidity of the ambient air. They consist of two thermometers, each inside a double tube for minimizing radiative heating both by the direct sun and longwave radiation exchange between thermometer and surrounding tube. The instrument is ventilated during measurement by a manually winded fan. One of the thermometers is covered by a cotton gaze that has to be wettened before the measurement. Depending on the ambient humidity, water is evaporating from the wet gaze and cooling down the wet thermometer that reaches the so called wet-bulb temperature after a few minutes. From the wet bulb temperature, the air temperature measured simultaneously by the dry thermometer and the air pressure, the humidity of the air can be calculated.

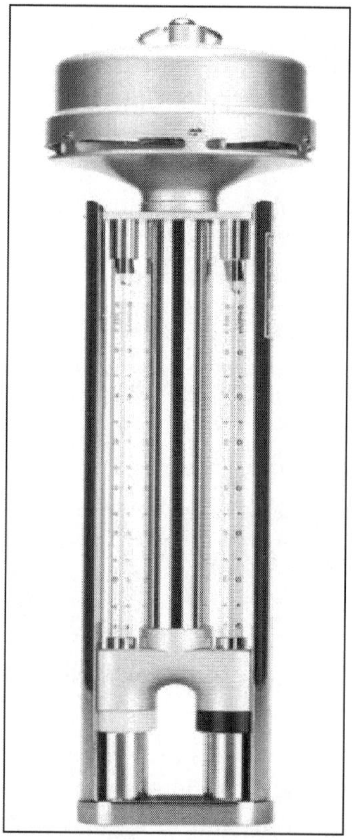

Figure 5.5. Assmann Psychrometer.

Whirling/Sling Psychrometer

It consists of dry bulb thermometer and wet bulb thermometer. It is used to measure temperature and relative humidity using hygrometric table in micrometeorology.

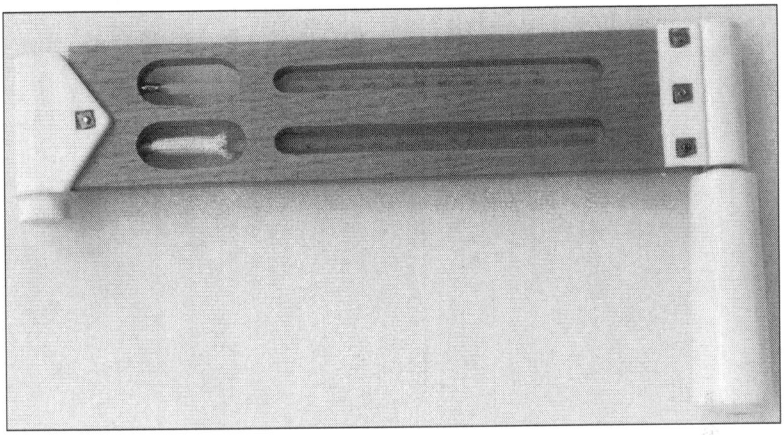

Figure 5.6. Whirling/Sling Psychrometer.

There are two thermometers in this instrument; Dry bulb thermometer: the sensor bulb is in direct contact with air (mercury is the sensing element) and wet bulb thermometer: the sensor bulb of this thermometer is covered with a wick of cotton or muslin moistened with pure water. A glass frame is provided as the housing for the instrument. A rotating handle is attached to the glass frame housing. In order to measure the dry bulb and the temperature of the wet bulb, the disposition of the thermometer of the psychrometer glass cover is rotated from 5-10 m/s to obtain the necessary air movement.

Infra-red Thermometer

It is a portable and easy to use instruments, suitable for field measurements. Infrared thermometer consists of a lens to focus the infrared (IR) energy on to a detector, which converts the energy to an electrical signal that can be displayed in units of temperature after being compensated for ambient temperature variation. This configuration facilitates temperature measurement from a distance without contact with the object to be measured. It is based on the Stefen-Boltzman radiation law which states that the energy flux density of radiation from an object is directly proportional to the fourth power of its absolute temperature. Thus, radiated energy coming from the object is received by the sensor and is converted into temperature. The air temperature is measured with help of mercury in the glass type thermometer but the accuracy of these thermometer is +/-0.5°C and response time is very large. Therefore, now air temperature is measured with the help of thermocouple and thermistor, wherein accuracy is +/-0.1°C and response time is very less. Micrometeorological measurements, air temperature is measured by thermocouple and canopy temperature is measured by infrared thermometers. Infrared thermometers are measured temperature by detecting the intensity of IR radiations emitted by an object. The IR radiations emitted by the object are received by the sensor which finally converts it into the target temperature.

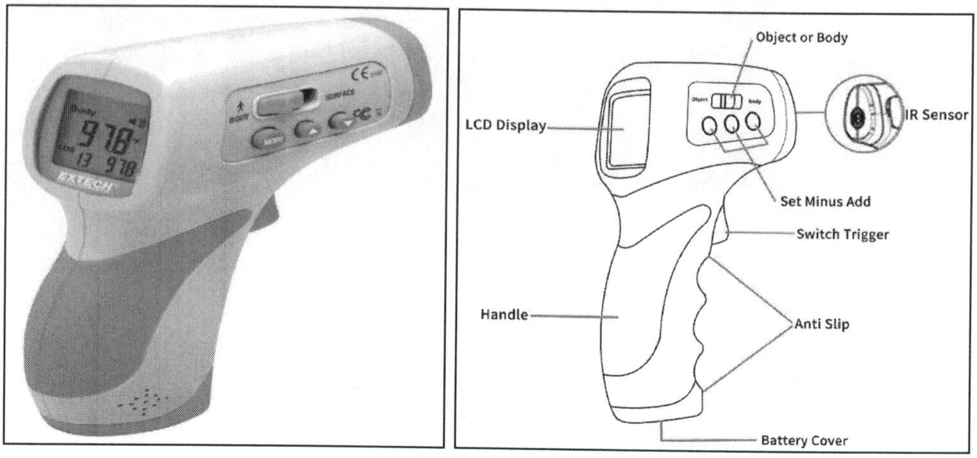

Figure 5.7. Infra-red Thermometer.

5.10 Soil Temperature

It is the measure of degree of hotness or coldness of a soil. It is express the intensity of heat energy available in response to the absorption of solar radiation directly and indirectly. It is important for plant growth and development, germination of seeds, soil microorganisms' activities, nitrification process, organic matter decomposition, mineralization, nutrient uptake *etc.* Dark colored soils absorb more heat and becomes warmer than the light soil during bright days. Microbial activities of nitrobacteria and heterotrophic organisms increases with rise in temperature. Optimum soil temperature range is 18-30°C. Soil acts as sink during day time and as a source of heat during night time.

5.10.1 Thermal Properties of Soils

- **Specific heat or mass specific heat:** It is the amount of heat required to raise the temperature of one gram of substance by 1°C.
- **Heat capacity or volume specific heat:** It is the amount of heat required to raise the temperature of 1 cubic centimeter substance by 1°C.
- **Thermal conductivity:** It is the quantity of heat transmitted through unit length of substance per unit cross section per unit temperature gradient per unit time.
- **Molecular conductivity** is the ability of the substance to transfer heat from molecules to molecules. It is expressed in J $m^{-1}s^{-1}k^{-1}$.

- ☆ **Temperature gradient** is the rate of change of temperature with displacement in a given direction.
- ☆ **Thermal diffusivity:** It is the ratio of the thermal conductivity to volume specific heat.
- ☆ **Soil heat flux:** It is the quantity of heat flowing into or out of unit area of soil in unit time and depends on temperature gradient and conductivity.

5.10.2 Variation of Soil Temperature

The daily change in soil temperature is due to changes in incoming solar radiation and outgoing terrestrial radiations. Maximum or minimum soil temperature takes time to reach lower depths know as thermal lag. Thermal lag increases with increase soil depth. At 40 cm depth the change is very minor and at 80 cm depth no daily change occurs. The daily maximum soil temperature depends on the texture of the soil *i.e.* sand > loam > clay soil. The soil surface is the coldest level in early morning and warmest in early afternoon. The greatest amount of heat is absorbed by the soil in summer season than winter season. The rate of penetration of heat wave within the soil takes around three hours to reach 10 cm depth.

5.10.3 Effect of Soil Temperature

- ☆ **Bioactivity:** Increase in soil temperature increases the soil nitrogen mineralization rates through the increase in microbial activity and increase in the decomposition of organic matter in the soil. Soil temperature below freezing point decreases mineralization by inhibiting microbial activity and decreasing diffusion of soluble substrates in the soil.
- ☆ **Soil aeration:** Temperature influences the carbon dioxide content in the soil air. High temperature encourages micro-organism activity which results in higher production of carbon dioxide in the soil.
- ☆ **Soil microorganisms:** Soil microbial activities decrease with low soil temperatures and at freezing point, most activities cease. At a soil temperature range of 10°C-24°C, soil macro-organisms have increased rate of metabolism requiring them to either feed more or burn their own fat stores.
- ☆ **Organic matter decomposition:** Organic matter decomposes slowly at lower temperatures as a result of decrease in microbial activities and biochemical processes. Soil temperature between 20-38°C increases the organic matter decomposition by increasing the movement of soluble substrates in the soil and stimulating microbial activities.
- ☆ **Water uptake:** Water uptake decreases with low temperature. This is due to the increased viscosity and decreased absorption rate of water at low temperature. Decreased water uptake reduces the rate of photosynthesis.

- **Cation exchange capacity:** Increase in soil temperature decreases organic matter through combustion. This decrease in organic matter and reduction in clay size clay fraction as a result of high temperature leads to a decrease in the cation exchange capacity of the soil.
- **Soil pH:** At a soil temperature ranges of 25°C-39°C the soil pH increases as a result of organic acid denaturation which increases at high temperature.
- **Soil moisture content:** Reductions in soil moisture occur when increased soil temperatures decrease water viscosity, thus allowing more water to percolate through the soil profile.
- **Root growth:** Increase in soil temperature improves root growth because of the increase in metabolic activity of root cells and the development of lateral roots.
- **Nutrient uptake:** The metabolic activities of micro-organisms play an important role in the cycling of nutrients in the soil and ensuring the nutrients are in a form available to plants. Therefore, increased metabolic activities of micro-organisms as a result of increase in soil temperature will stimulate the availability of nutrients for plants.

5.10.4 Factors Affecting the Soil Temperature

- **Evaporation:** The greater the rate of evaporation the more the soil is cooled.
- **Rainfall:** Rainfall cools down the soil.
- **Condensation:** Water vapour from soil, condensed in the air, its heat increases noticeably.
- **Solar radiation:** The amount of solar radiation available at any given location and point of time is directly proportional to soil temperature.
- **Vegetation:** Bare soil absorb heat and become very hot during summer and cold during the winter.
- **Colour of the soil:** Black colour soil absorbs more heat than light coloured soil.
- **Moisture content:** A soil with higher moisture content is cooler than dry soil.
- **Tillage:** The cultivated soil has greater temperature than uncultivated soil.
- **Soil texture:** Thermal conductivity decrease with reduction in particles size.
- **Organic matter content:** OM reduce the heat capacity and thermal conductivity of soil.
- **Slope of land:** The amount of solar radiation reaching per unit area of the land surface decrease as the slope of the land is increases.

5.10.5 Instruments Used to Measure Soil Temperature

Soil Thermometer

The soil thermometer is mercury in glass type thermometer. The scale is enclosed inside the thermometer. Soil thermometer is installed in the observatory at the proper recommended site on a small plot of size 2.5 m by 1.5 m.

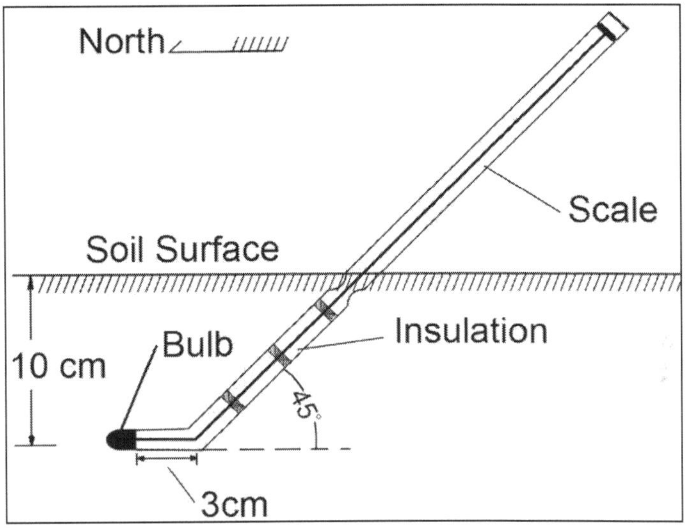

Figure 5.8. Soil Thermometer.

The range of soil thermometer is -20°C to +60°C. These thermometers with increasing depth are fixed from west to east, *i.e.* soil thermometer for 5 cm, depth is in the west followed by the thermometers of the depth of 10 cm, 20 cm. Soil thermometers for 3 depths, *i.e.* 5 cm, 10 cm and 20 cm are placed 45 cm apart at an inclined depth of 5.8, 11.6 and 23.2 cm to ensure a vertical depth of 5, 10 and 20 cm, respectively. These thermometers have a bend bulb of 120° and the rest of the stem is straight. These thermometers are held by the iron stands in inclined position making 60° angles with ground surface. The thermometer is installed so that they face towards south. The diurnal range of soil temperature is highest at the surface and this range decreases rapidly with depths and becomes practically negligible at a depth of 30 cm. In the morning the temperature is lowest at 5 cm depth and in the afternoon, it is highest at 5 cm depth. The distance between the ground and the 5°C on the thermometer scale is called check distances. This distance at the time of installation is measured and must be maintained always.

Installation

- ✰ The site for such measurements should be a level plot of size 180 cm x 120 cm in the observatory and typical of the surrounding soil for which information is required.

- Minimum soil layer should be disturbed during installation of thermometers inside the soil.
- The bulb of the thermometer should rest in good contact with the firm undisturbed soil.
- While digging the soil for installation, the soil should be removed layer wise and identically layers put in corresponding previous position.
- The zero mark of the thermometer should be at ground level

Recording Observations

- The soil temperature readings are taken read daily at 07.00 and 14.00 hr LMT.
- The following points should be kept in mind while taking the observations.
- The line joining the observer's eye and top of the mercury column should be at right angle to the instrument.
- The observer should stand at some distance from the instrument so as not to shadow the ground surface in the vicinity of the instrument.
- Reading should be taken correct to 0.10°C.

Chapter 6

Solar Radiation

6.1 Introduction

Radiations is the transfer of thermal energy in the form of electromagnetic waves from one place to another through vacuum with speed of light (3×10^8 m/sec). Solar radiation is the solar energy that reaches the earth surface in the form of electromagnetic wave. Solar radiation is the form of energy that is received from sun in the form of insolation through short wave length. Insolation is the incoming solar radiation or energy received on the surface of the earth. Incoming solar radiation is the solar radiation that comes to the earth. Outgoing radiation is the solar radiation that leaves the earth surface. Isohel is the line joining the places having equal duration of sunshine hours.

6.2 Important Definition about Radiation

Short wave radiation: Radiation having the wave length less than 4 μm is called as short-wave radiation. Incoming solar radiation is short wave radiation. It is measured with the help of pyranometer.

Long wave radiation: Radiation having the wave length longer than 4 μm is called as long wave radiation. Outgoing radiation is long wave radiation. It is measured with the help of pyrgeometer.

Diffuse sky radiation is solar radiation reaching the earth surface after having been scattered from the direct solar beam by molecules or particulates in the atmosphere.

Atmospheric window: It is long wave radiation escapes to the space between 8.5 and 11μ. This act as transparent medium instead of absorbing. It is used in microwave remote sensing to monitor the features of the sky in case of overcast sky.

Terrestrial radiation is the radiation emitted both from the earth and atmosphere. Thermal radiation is electromagnetic radiation generated by the thermal motion of particles in matter.

Direct radiation is the direct beam solar radiation received by a unit horizontal surface. It is measured with the help of pyrheliometer. Net radiation is the difference between total incoming radiation and outgoing radiations. It is measured with the help of solarimeter or net pyranometer or net pyrradiometer.

Global radiation is the total of direct beam solar radiation and the diffuse sky radiation received by a unit horizontal surface. It is measured with the help of pyrradiometer.

Counter radiation is the longwave radiation radiated by the atmosphere back to the earth surface. This radiation prevents the earth from excessive cooling.

Absorptivity is the ratio of the amount of radiant energy absorbed to the total amount incident upon that substance.

Reflectivity is the ratio of the radiant energy reflected to the total incident radiation upon that surface. If it is expressed in percentage, it is known as albedo.

Transmittivity is the ratio of the transmitted radiation to the total incident radiation upon the surface.

The heat flow in unit time is known as flux.

Emissivity is the ratio of the radiant energy emitted by a given surface to the total heat energy emitted by a black body. The emissivity of a black body is unity.

Blackbody radiation is a body, which completely absorbs all the heat radiations falling on it without reflecting and transmitting any of it *i.e.* reflectivity and transmittivity become zero.

Radiant flux is the amount of energy radiated, transmitted or absorbed per unit time. Its unit is watt.

Metabolic Heat is the part of radiation which is utilized by the plants in respiration and in maintaining its canopy temperature.

Radiant flux density or radiation intensity is defined as radiant flux per unit area. Its unit is watt/m^2.

Irradiance is the radiation intensity incident on a surface.

Emittance is the radiant flux density emitted by a surface.

Radiance is the ratio of radiation intensity and apparent area.

6.3 Solar Constant

Solar constant is the radiation energy per unit time per unit area on surface, which is held perpendicular to sun rays at outer limit of atmosphere at mean distance between sun and earth. It is the average amount of solar radiation reaching the outer limits of the atmosphere. This occurs in April and October. The Sun emits radiation about 56 × 10^{26} cal/minute. The mean distance between sun and earth is 1.5 × 10^{13} cm. The mean value of solar constant is 1.98 cal/cm^2/min or 1398 watt/m^2 or 2.00 ly (Langley's).

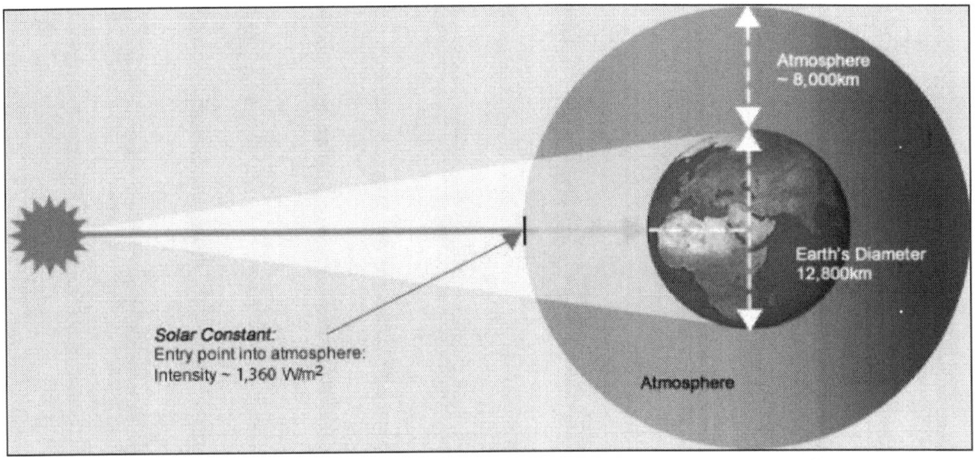

Solar constant = Total radiation/Area of spherical cell

$= 56 \times 10^{26}$ cal $min^{-1}/4\pi r^2$ (r = mean distance between sun and earth)

$= 56 \times 10^{26}$ cal $min^{-1}/4\pi\ (1.5 \times 10^{13})^2$

$= 1.98$ cal $cm^{-1}min^{-1}$

6.4 Albedo

It is the ratio of reflected radiation to the incident radiation and expresses in percentage. It is derived from Latin word albedo or whiteness means reflecting power of a surface. It is varying with colour and composition of the earth's surface, season, angle of the sun rays). Value is highest in winter and at sunrise and sunset.

$$\text{Albedo} = \frac{\text{Reflected solar radiation by a surface}}{\text{Total solar radiation incident on the surface}} \times 100$$

The values of albedo for fresh snow (80-85 per cent), clouds (55 per cent), earth (34-43 per cent), grasses (20-25 per cent), soils (15-50 per cent), grain crops (15-25 per cent), vegetation (10-40 per cent) and pure water (5-20 per cent).

6.5 Laws of Radiation

a) Stefan-Boltzmann Law

The radiation intensity emitted by a black body is directly proportional to the fourth power of its surface temperature.

$$E = \varepsilon \sigma T^4$$

where,

E = Radiation emitted by the surface

σ = Stefan-Boltzmann constant (8.14×10^{-11} ly $min^{-1}K^{-4}$)

T = Surface temperature (°K)

ε = Emissivity

b) Planks Law

Energy content of a quantum is directly proportional to its frequency.

$$E = h \nu$$

where,

E = Energy

v = Frequency

h = Planks constant (6.626×10^{-34} J **sec^{-1}**)

c) Wien's Displacement Law

The wavelength of maximum radiation intensity emitted is inversely proportional to the temperature of the body.

$$\lambda_{max} = w/T$$

where,

λ_{max} = Wavelength of maximum emission (μm)

T = Temperature of the surface (°K)

w = Weins constant = 2897

d) Kirchoff Law

The absorptivity of a body for radiation is equal to its emissivity for the same wavelength.

$$a(\lambda) = e(\lambda)$$

where,

a = Absorptivity

e = Emissivity

λ = Wavelength

e) Lamberts Cosine Law

The radiant intensity emitted in any direction from a unit radiating surface or incident on a surface varies as the cosine of the angle between the normal to the surface and direction of the radiation.

$$I/I_0 = \text{Cos}\ \alpha$$

where,

I = Radiation intensity on the surface

Io = Radiation intensity when the solar radiation is normal to surface

α = Angle between radiation and normal to the surface.

f Beer's Law

The fraction of radiation intercepted by the canopy is an exponential function of leaf area index.

$$I/I_0 = e^{-KF}$$

where,

I = Light intensity at a level inside the canopy

I_0 = Light intensity at the top of the crop canopy

K = Extinction coefficient

F = Leaf area index from top to the level inside the canopy at which RI is to be calculated

g) Monteith Law

Radiation interception inside the crop canopy is expressed as by this law;

$$I = I_0[S+(1-S)\tau]F$$

where,

S = Fraction of light passing through unit layer without interception

τ = Leaf transmission coefficient

F = Leaf area index

6.6 Solar Radiation Spectrum

The range of wavelengths of solar radiation is called as solar radiation spectrum. or it is part of the electromagnetic spectrum of radiant energy that includes ultraviolet, visible and infra-red wave lengths.

Solar Spectrum		Spectral Range	Characteristics
Ultraviolet (UV)(9 per cent)	UV_A	0.32 to 0.40 µm	It can cause some damage to living cells. It is transmitted to earth surface and not absorbed in stratosphere.
	UV_B	0.28 to 0.32 µm	It is absorbed in stratosphere by ozone. It causes damage to living cells.
	UV_C	0.15 to 0.28 µm	It is completely absorbed by ozone. It is lethal and kills the living cell.
Visible (PAR)(41 per cent)	Blue	0.4 to 0.5 µm	It is the strongest chlorophyll and yellow pigment absorption region. The photosynthetic activity becomes very strong and effect on formation of tissues.
	Green	0.5 to 0.6 µm	It has low photosynthetic and photo periodic activity.
	Red	0.6 to 0.7 µm	It generates strong photosynthetic and photo periodic activity and light is strongly absorbed by chlorophylls.
Infrared (IR)(50 per cent)	Near IR	0.7 to 1.3 µm	It has important role on photoperiodism, germination of seeds, flowering and coloration of fruits. It is helps in plant elongation.
	Middle IR	1.3 to 3.0 µm	No specific effect on plant activity. Radiation absorbed by plants are transformed into heat. This radiation does not interfere with bio chemical processes.
	Far IR	3.0 to 30 µm	

6.7 Factors affecting Distribution of Solar Radiation

6.7.1 Factors Affecting Distribution of Solar Radiation Over the Earth Surface

- **Angle of incidence:** Radiation intensity is more when the sunrays are normal to the surface as compared to the sun rays are oblique to the surface or as the incident angle of sun rays' decreases, radiation intensity increases.
- **Latitude:** With increase in latitude the radiation intensity decreases. Therefore, radiation intensity is higher over equator as compared to poles.
- **Duration of sunshine:** The longer periods of sunshine supply more radiation to the surface as compared to shorter sunshine periods.

- **Distance between sun and earth:** The distance between earth and sun varies because of elliptical orbit of revolution of earth around sun. The earth comes closer to sun at perihelion and goes little farther from sun at aphelion. Therefore, the radiation intensity is 7 per cent greater at perihelion than at aphelion.
- **Atmospheric condition:** Reflection and absorption of radiation by clouds and other atmospheric constituents control the radiation intensity reaching the earth surface. Therefore, areas with cloudy atmosphere receive less radiation as compared to areas with clear and less turbid atmosphere.
- **Thickness of atmosphere:** With the increase in thickness of atmosphere the radiation has to pass more path in the atmosphere. Therefore, as the path length increases, the depletion of radiation increases or the atmospheric absorption of radiation increases with the thickness of atmosphere, thereby, less radiation reaches the earth surface.
- **Altitude**: The flux of solar radiations received increase with increasing altitude. Mountain areas receive more radiation while the valley floor may receive less amount.

6.7.2 Factors Affecting Distribution of Solar Radiation within the Plant Canopy

- **Type of plants:** The leaves of cereals crops have a Transmissivity from 5-10 per cent. The broad leaves of ever green plants have lower value of 2 to 8 per cent, whereas aquatic plants have 4 to 8 per cent.
- **Age of the leaves:** The transmissivity of younger leaves is more as compared to older leaves.
- **Chlorophyll content:** As the chlorophyll content increase the values of transmissivity decrease.
- **Arrangement of leaves:** The relative light interception by horizontal and erect foliage is 1:0.44. When the leaf area index is one the light transmissivity of more upright leaves is 74 per cent as against 50 per cent for horizontal leaves.
- **Angle of leaves:** In full sunlight, the optimum inclination for efficient light use is **81°**. The ideal arrangements of leaves are that the lowest 13 per cent of the leaves lay at angles between 0-30° to the horizontal, that the adjoining 37 per cent of the leaves lay at 30-60° and the upper 50 per cent leaves lay at 60-90°.
- **Plant height**: When the plant height increases the Transmissivity of light by the canopy decrease.
- **Angle of the sun:** The highest radiation penetration occurs at noon. Relatively high radiation penetration also occurs both in the morning as well as before the sunset due to high proportion of diffuse light.

☆ **Plant density:** In sparse crop stands the percent of light transmissivity is more and less in dense crop canopies.

6.8 Transfer of Heat

a) Conduction

It is the transfer of heat by molecular activity. The transfer of heat by matter through its adjacent layer from its hotter to colder parts without any motion of the matter itself. Heat transfer through matter without the actual movement of the substances or matter. Heat flows from the warmer to cooler part of the body so that the temperatures between them are equalized. *e.g.* The energy transmission through an iron rod which is made warmer at one end. Heat is received when hot objects are touched while lost with cold ones.

b) Convection

The transfer of heat through the mass movement of a hot fluid matter either liquid or gas is known as convection. Processes of transmission of heat through the actual movement of molecules of the medium. This is predominant form of energy transmission on the earth as in all the weather-related processes. *e.g.* Boiling of water in a beaker.

c) Radiation

The transfer of heat from a hot body to cold without any material medium is known as radiation. Transfer of energy from one body to another without the aid of the material medium (solid, liquid or gas). Radiation is not heat, only when radiation is absorbed by surface of a body heat is produced. *e.g.* The energy transmission through space from the sun to the earth.

6.9 Energy Balance of Earth

The difference between all incoming and outgoing radiation at the earth surface and top of the atmosphere is known as radiation balance of the earth surface. Soil heat flux is the flow of heat into or out of the soil through a process of conduction. Sensible heat flux is the flow of heat into or out of the soil through a process of convection. Latent heat flux is the flow of heat into or out of the soil through a process of condensation. The basic form of radiation balance equation for the earth and its atmosphere is:

$$\mathbf{Rn = G + H + LE + PS + M}$$

where,

Rn = Radiation balance/net radiation

G = Surface soil heat flux

H = Sensible heat flux

LE = Evapotranspiration loss

PS = Photosynthesis loss

M = Miscellaneous loss

If the incoming solar radiation (insolation) is considered as 100 unit at the outer atmosphere, 30 units are returned to space through reflection from the earth surface (4 unit), Clouds (20 units) and back scattered by air (6 unit) without heating the atmosphere. Out of 70 units remaining, 16 units are absorbed by water vapour, dust and ozone and 3 unit are absorbed by Cloud. Out of 51 units absorbed at the earth surface, 23 units are used for latent heat flux and 7 units are used for sensible heat flux and the earth emits 21 units in the form of long wave radiation. Out of 21 units emitted by earth, 15 units are absorbed by water vapour and carbon dioxide and 6 units are lost to space.

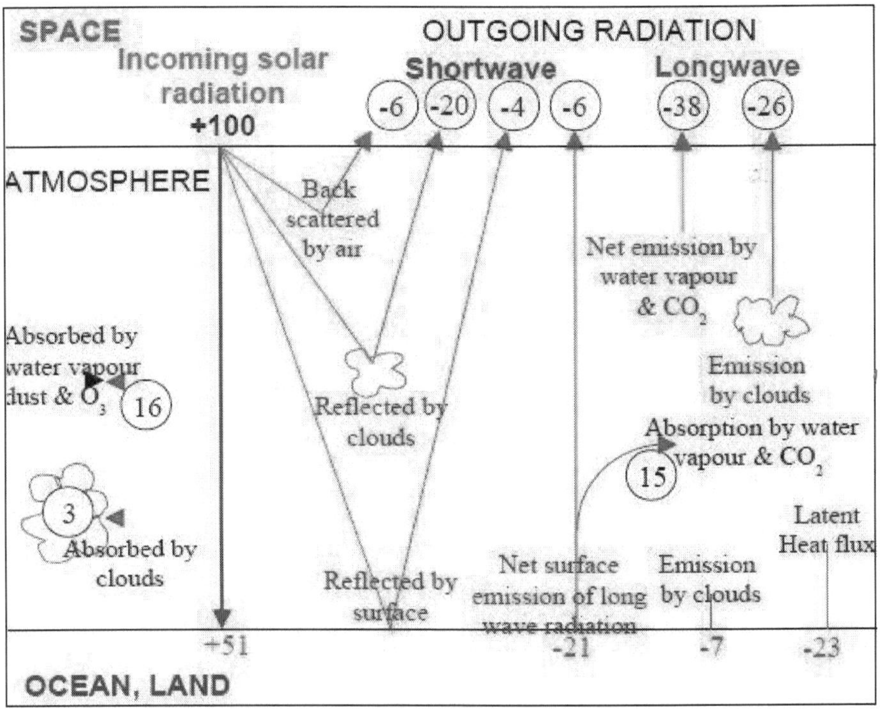

Figure 6.1. Energy Balance of Earth.

6.10 Important Facts: Scattering of Sun Rays

Scattering is the process by which molecules of a medium and suspended small particles in the medium diffuse a portion of the incident radiation in all direction.

a) Rayleigh Scattering

It occurs when particles (dust, smoke, water droplets, N_2, O_2 molecules) are very small compared to the wavelength or radiation. The amount of scatter for particles whose radius is less than 1/10th of wavelength of incident light is inversely proportional to the fourth power of wavelength. *e.g.* blue color of the sky, red and orange colours at sunrise and sunset are due to Rayleigh scattering at upper atmosphere. Blue colour of sky is due to scattering of visible solar spectrum by gas molecules and dust particles. Red Colour of the sky at sunset and sunrise is because of increased path length in the atmosphere. Percentage of solar energy in the visible part decreases. With in the visible part, the ratio of the blue to the red part decreases with increased path length.

b) Mie Scattering

It occurs when the atmospheric particles are about the same size of the wavelength of the radiation. For larger particles with circumference >30 times of wavelength of the incident radiation, scattering is independent of the wavelength (*i.e*) white light is scattered.

c) Non-Selective Scattering

It occurs when the particles sizes are larger than the wavelength of radiation. *e.g.* whitish appearance of fog and clouds.

6.11. Instruments Used to Measure Radiations

a) Campbell Stokes Sunshine Recorder

It is used to measure the duration of bright sunshine hours. The spherical glass Lense is well annealed colourless optical glass. The sun rays are focused by a convex glass sphere to an intense spot which as the moves across the sky, burn a trace in a card placed at the focal length of the glass sphere. The length of the burns or trace is proportional to the duration of bright sunshine hours. The long-curved cards, short curved cards and straight cards are placed in concentric groves in the bowl.

Figure 6.2. Sunshine Recorder.

The instrument consists of a glass sphere of 10 cm diameter, mounted concentrically in a section of a spherical bowl. The instrument is installed on a masonry pillar of 10 ft (3.04 m) above ground. Three overlapping pairs of grooves are provided in the bowl to insert chemically sensitized cards suitable for different seasons of the year.

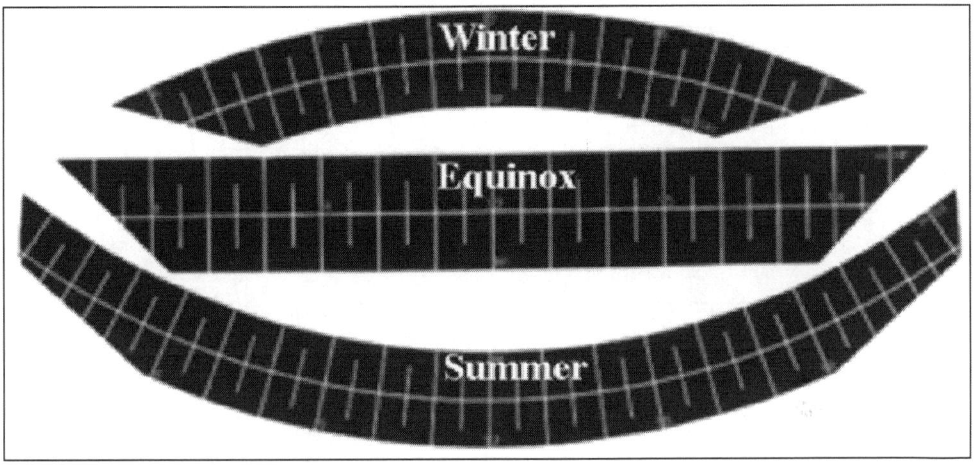

Figure 6.3. Sunshine Cards.

Short curve cards are inserted on the upper groove in winter season from 15th October to end of February. Long curve cards are inserted on the lower groove in summer season from 12th April to 2nd September. Straight cards are inserted on the middle groove during equinoxes from 3rd September to 14th October and 1st March to 11th April. The new card is inserted in the appropriate groove before sunrise or after sunset and removed the burnt card in the evening after sunset. The action of the recorder depends upon the burning of the card due to heat of the sun, which are focused on the card through the glass sphere. The cards burn linearly because some chemical treatment is given to the cards. The total length of the burn is measured with the help of a time scale to obtain the duration of sunshine.

b) Pyranometers

It is used to measure short wave radiation *i.e.* direct and indirect radiation. It is a thermopile-based instrument.

c) Pyrheliometers

It is used to measure direct component of solar radiation at normal incidence.

d) Pyrradiometers

It is used to measure short wave radiation and long wave radiation.

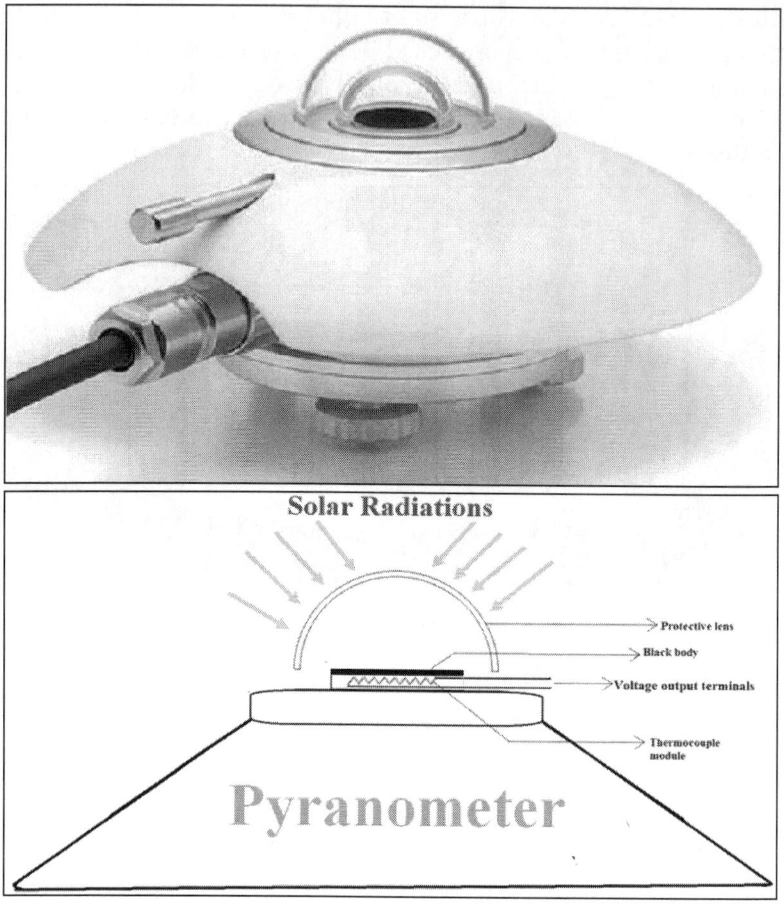

Figure 6.4. Pyranometer.

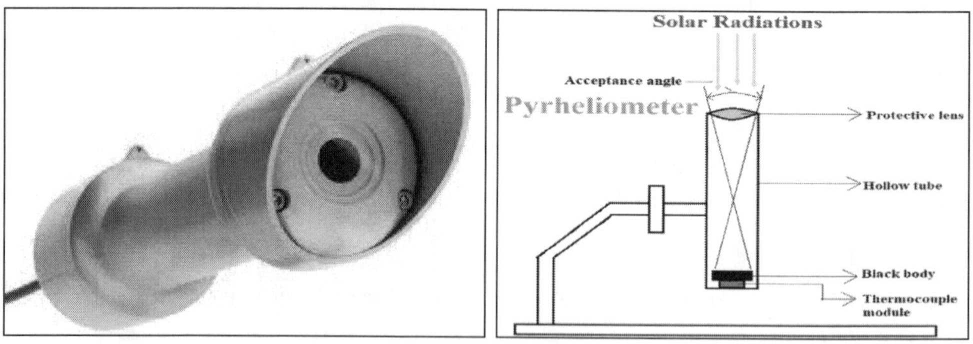

Figure 6.5. Pyrheliometer.

e) Net Radiometer

It is used to measure net radiation. A differential thermopile is used in this radiometer. It measures net radiation flux over the surface.

f) Pyrgeometer

It is instrument used for measuring long wave or terrestrial or infrared radiations. The sensor used is a thermopile, which works on the principle of thermo-electric effect.

g) Line Quantum Sensor

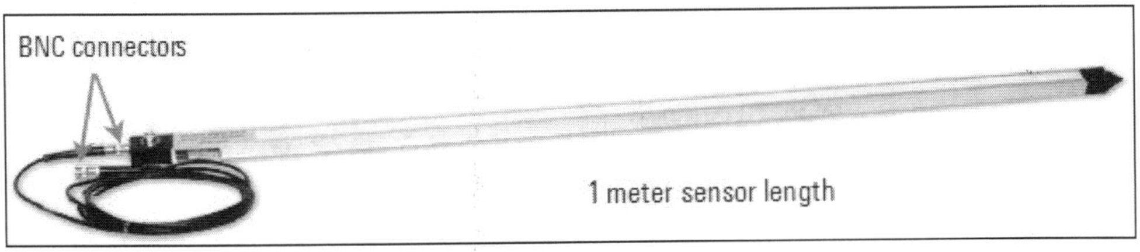

Figure 6.6. Line Quantum Sensor.

It is designed for measuring PAR in application where the radiation to be measured within plant canopies. The sensor is silicon photodiode, which works on the principle of photometric effect. The sensor has the quantum response through the wavelength range of 400 to 700 nm for photosynthetic photon flux density as generally preferred for PAR measurement and has an output in units of μ mols^{-1} m^{-2}. Line Quantum Sensor is usually a rod of 1 m length and 12.7 mm in width. A number of high stability silicone photo voltaic detectors are arranged linearly along its length at a distance of 2.38 cm from each other. A water proof anodized aluminum case encloses the detectors. The output from the sensor is connected to an integrating quantum meter along with a calibrated connector. The values are displayed on an LCD display. The output of a quantum sensor is given in u E s^{-1}m^{-2} (1 Wm^{-2} = 0.2174 u E s^{-1} m^{-2}). In order to measure the IPAR (Intercepted), the line quantum sensor is placed above the crop canopy or in the open. The sensor is inverted over the crop canopy to get RPAR (Reflected). For measuring TPAR, the sensor is placed perpendicular to the row direction of the crop horizontally at the ground level. APR is then estimated as:

$$APAR = IPAR - (RPAR + TPAR)$$

The values are expressed in terms of percent of IPAR. For the measurement of profile APAR, the sensor is placed at 25, 50 and 75 per cent of the crop height and TPAR is recorded at the respective heights. It is generally advisable to take readings between 1100 and 1300 hr IST for one reading per day. In case of diurnal observation requirements, hourly readings should be noted from sunrise to sunset.

i) Luxmeter

It is used to measure light intensity of visible radiation that can be sensed by human. It measures illuminance of a surface by the incident light and its unit is lux.

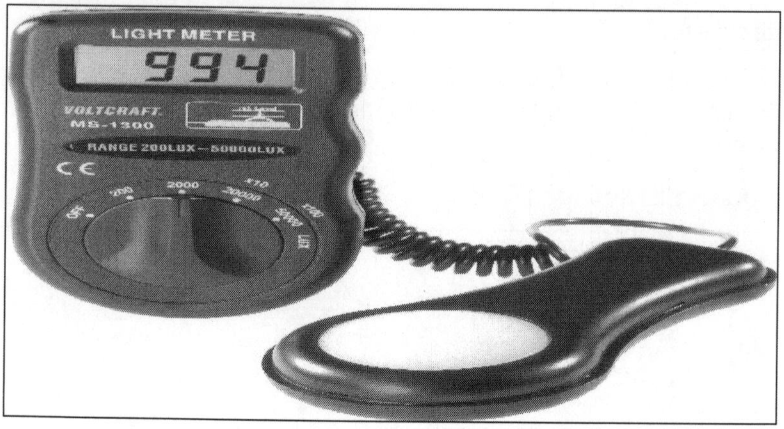

Figure 6.7. Luxmeter.

6.12 Growing Degree Day

The GDD also called heat unit, effective heat unit, growth unit are a simple means of relating plant growth, development and maturity to air temperature. Reaumer introduced the concept of heat unity. The GDD concept assures a direct and linear relationship between growth and temperature. It starts with the assumption that the growth of a plant is dependent on the total amount of heat to which it is subjected during its lifetime. A degree day is the departure from the mean daily temperature above the minimum threshold temperature or base temperature. The minimum threshold is the temperature below which no growth takes place. The threshold varies with different plants, and for majority of plants it ranges from 4.5 to 12.5°C, with higher values for tropical plants and lower values for temperate plants.

Listener's law states that the total heat unit for a crop, varies from region to region and variety to variety but the ratio of the total growth heat units to the heat units required to produce a given phase of a growth in a given crop in a given region is supposed to be nearly constant.

$$GDD = T_{max} + T_{min}/2 - T_{base}$$

where,

T_{max} = Maximum Temperature

T_{min} = Minimum Temperature

T_{base} = Base Temperature

Name of Crop	*Base Temperature (°C)*	*Name of Crop*	*Base Temperature (°C)*
Rice	5	Cotton	14
Millets	8	Coconut	13
Maize	10	Sorghum	10
Wheat	4.5		

Merit and Application

- ✰ It is useful in agricultural operations and planning. Forecast of the appearance of phonological stages, maturity, crop harvest, yield and quality can be possible by studying heat units required for such stages.
- ✰ It is useful to the problems of growth and development of plants due to pathogens, animals, insects.
- ✰ Possibility of introduction of new crop in a area in which it has not been grown earlier can be determined.
- ✰ More efficient variety from many varieties suited to the original climate can be selected.
- ✰ Microclimate modification in such a way as to the optimum for each developmental stage of an organisms can be decided.

Demerits

- ✰ The theory uses earlier basis of temperature for getting growth. Plant activity or growth does not take place at equal rate at different temperatures.
- ✰ Use of same threshold or base temperature for every advancing temperature and the use of Shade temperatures in Stevenson screen divided from the actual on site or in crop temperature.

6.13 Agro Meteorological Indices

Photo Thermal Units (PTU)

It is the product of the growing degree day and the length of the day accumulated over a given period and expressed as °C day hrs.

Helio Thermal Units (HTU)

It is the product of the growing degree day and actual bright sunshine hours and expressed as °C day hrs.

Hydro Thermal Unit (HgTU)

It is the product of the growing degree day and the relative humidity [morning (RH-I) and evening (RH-II)] and expressed as °C percent.

Photo Thermal Use Efficiency

It is the amount of dry matter or grain yield produced per unit of accumulated photo-thermal units and expressed as kg ha^{-1} °C day^{-1} hrs^{-1}.

Helio Thermal Use Efficiency

It is the amount of dry matter or grain yield produced per unit of accumulated helio-thermal units and expressed as kg ha^{-1} °C day^{-1} hrs^{-1}.

Hygro Thermal Units Use Efficiency

It is the ratio of yield (kg ha^{-1}) and hydro thermal unit (°C day) and expressed as kg ha^{-1} °C day^{-1} hrs^{-1} percent.

Heat Use Efficiency (HUE)

It is the amount of dry matter yield produced per unit of growing degree days or thermal time. It is also represented by thermal use efficiency and expressed as kg ha^{-1} °C day^{-1} hrs^{-1}.

Thermal Interceptions Rate (TIR)

It is the multiplication of heat units with solar radiation intercepted by a plant and expressed as °C day^{-1} MJ $plant^{-1}$.

Radiation Use Efficiency (RUE)

It is introduced by Monteith showed that the accumulated dry matter production of a crop was linearly related to accumulated intercepted solar radiation. It is the quantity of biomass produced per unit of intercepted radiation and expressed as °C day^{-1} MJ m^{-1}.

Moisture Availability Index (MAI)

It is the ratio of rainfall at 75 per cent probability and PET and given by Hargreaves.

Moisture Adequacy Index (MAI)

It is the ratio of actual evapotranspiration and potential evapotranspiration.

MAI	*Category*	*Moisture Status*
> 75 per cent	Excellent	Adequate moisture
50 - 74 per cent	Good	
24 - 49 per cent	Poor	Moisture Stress
< 24 per cent	Very Poor	

Water Requirement Satisfaction Index (WRSI)

It is an indicator of crop performance based on the availability of water to the crop during growing season and for growing season is calculated as the seasonal AET to the seasonal crop water requirements. WR of the crop at a given time of the growing season is calculated by multiplying the reference ET with crop coefficient. Whose value are published by FAO.

Chapter 7

Atmosphere Humidity

7.1 Introduction

The amount of water vapour present in atmosphere is known as atmospheric humidity. The water in gas form in the atmosphere is known as water vapour.

7.1.1 Important Terminology of Humidity

- Vapour pressure is the pressure exerted by water vapour in air on the earth surface. It is expressed in mb or kpal.
- Actual vapour pressure is the pressure exerted by actually present in air (dry air).
- Saturated vapour pressure is the pressure exerted by saturated air.
- Saturation deficit/vapour pressure deficit is the difference between saturated vapour pressure and actual vapour pressure.
- Absolute humidity is the ratio of mass of water vapour to the volume of air. It is expressed in g/cm^3.

$$AH = \frac{\text{Mass of water vapour}}{\text{Volume of air}}$$

- Specific humidity is the ratio mass of water vapour to the mass of moist air. It is expressed in gm/kg.

$$SH = x\frac{\text{mass of water vapour}}{\text{mass of moist air}}$$

- Relative humidity is the ratio of actual vapour pressure and saturated vapour pressure and expressed in per cent.

$$RH = \frac{\text{actual vapour pressure}}{\text{saturated vapour pressure}} \times 100$$

7.2 Variation in Relative Humidity

Daily Variation in Relative Humidity

The daily variation of relative humidity is inverse to the temperature. At sunrise the relative humidity is maximum and minimum between 1400 to 1500 hrs.

Annual Variation in Relative Humidity

The annual variation of relative humidity is largely depending upon the locality. The maximum relative humidity occurs in winter season and minimum in summer season. Over ocean relative humidity reaches maximum in summer.

Variation of Relative Humidity with Latitude

Relative humidity shows maximum at equator, it decreases in regions of high-pressure belt and afterwards increase again in the polar region.

Variation of Relative Humidity with Altitude

In near cloud, relative humidity is 100 per cent but below or above it is different.

7.3 Factors affecting Relative Humidity

- **Water vapour content:** Humidity increases with increase in its water vapour content at the same temperature. Specific humidity is higher in tropics as compared to polar region.
- **Temperature of air:** The capacity of air to hold water vapour will depend upon its temperature. As the temperature of a given air sample increases, its humidity decreases. So, water vapour holding capacity of air increases with increase in temperature.
- **Distance from water bodies:** The places near water bodies have high humidity as compared to far places. Therefore, humidity is higher in coastal region as compared to interior land regions.
- **Prevailing winds:** The humidity of air coming from dry places is lower as compared to air from wet/humid places.
- **Vegetation:** Presence of vegetation increases humidity through loss of water vapour in atmosphere by evaporation and transpiration. So, vegetated regions are more humid than non-vegetated regions.
- **Condition of atmosphere:** Cloudy day is more humid than clear day. Precipitation is main source of water over land surface. So, rainy weather is more humid than dry or clear weather.

7.4 Instruments to Measure Humidity

Sling Psychrometer

It is used to measure both the dry bulb and wet bulb temperatures at time. These temperatures are a measure of humidity content in air. There are two thermometers in this instrument; a. Dry bulb thermometer: the sensor bulb is in direct contact with air (mercury is the sensing element) and b. wet bulb thermometer: the sensor bulb of this thermometer is covered with a wick of cotton or muslin moistened with pure water. A glass frame is provided as the housing for the instrument. A rotating handle is attached to the glass frame housing. In order to measure the dry bulb and the temperature of the wet bulb, the disposition of the thermometer of the psychrometer glass cover is rotated from 5-10 m/s to obtain the necessary air movement. Reading from the both thermometer is used to calculate humidity.

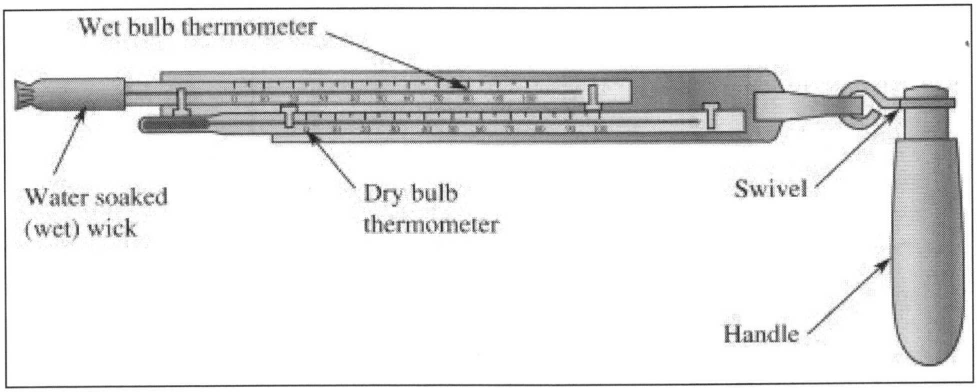

Figure 7.1. Sling Psychrometer.

Hygrograph

This is a self-recording device which can continuously record changes in humidity with time on the graph paper. It consists of a bundle of human hair. The changes in the length of hair due to changes in humidity are fed to lever mechanism which magnifies this micro motion, longer arm of the lever is connected with pen and ink which makes marking on the graph paper. Graph paper is calibrated in percent humidity on Y axis and in time in X axis. It is fixed on clock driven rotating drum.

Assmann's Psychrometer

This consist of two identical mercury in glass thermometer with rang -10°C to +60°C. They are mounted side by side in a nickel plated highly polished frame. Both the thermometer bulbs are placed in the coaxial metal tubes. Thermometer are insulated from the frame.

Figure 7.2. Hygrograph.

There is a clock driven exhaust fan which sucks air inside from the bottom. Air imparts heat to thermometer bulb and temperature are recorded.

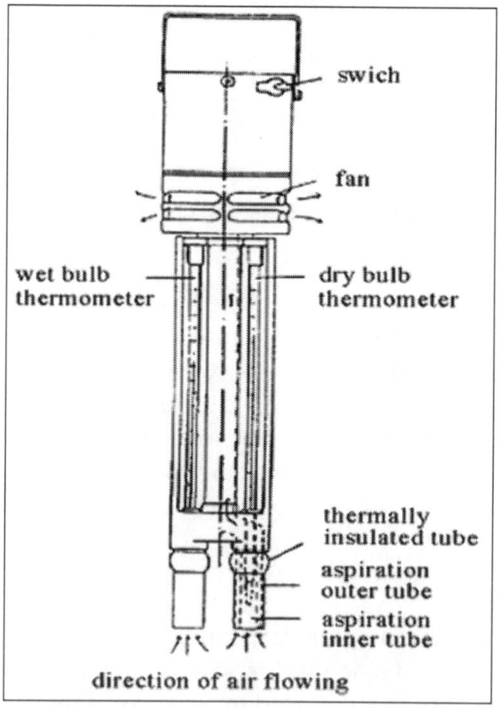

Figure 7.3 Assmann's Psychrometer

Chapter 8

Condensation

8.1 Introduction

Condensation is a process by which water vapor changes to liquid state. The air is cooled to a temperature at which condensation starts is called condensation temperature. The process by which water vapour directly changes to solid state is known as sublimation.

8.2 Conditions for Condensation

☆ **Presence of sufficient water vapour:** An adequate amount of water vapour is necessary to bring out saturation of air.

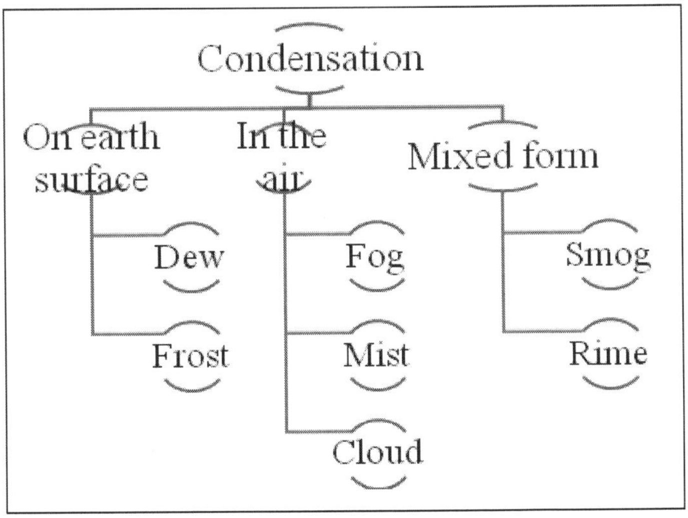

- ☆ **Saturation of air:** The process of condensation will take place only when air attains saturation level. It can be attained by adding water vapor in the air and cooling of the air.
- ☆ **Presence of condensation nuclei:** Sub microscopic or microscopic or particles like salts, combustion products released from industry, dust *etc.* act as nuclei for condensation. In the absence of hygroscopic nuclei, condensation cannot trigger even if the air is super saturated with relative humidity greater than 100 per cent and temperature below freezing point.
- ☆ **Cooling of air:** For saturation of atmospheric air with vapour it's cooling up to and below dew point is necessary and this cooling is takes place by adiabatic and non-adiabatic processes.

8.3 Forms of Condensation

8.3.1 Dew

It is the deposition of water vapour in the form of tiny droplets on the colder bodies on earth surface by condensation. It occurs in early morning. Dew point is the temperature to which air is cooled and air becomes saturated and condensation of water vapour starts. Clear sky, absence of wind, objects on which dew forms must be good radiators and bad conductor, water vapour content of air are conditions required for dew formation. Clouds reduce the radiational cooling during nights as they absorb terrestrial radiation and re-emit radiation towards earth surface. The cooling of surface objects by radiational cooling. During the night when the hot air comes into contact with some cold surface, water vapor present in it condenses on the cold surface in the form of droplets.

8.3.2 Frost

It is the deposition of water vapour in the form of ice crystals on the colder bodies on earth surface by condensation. Frost is frequently called as form of sublimation. Clear sky, absence of wind, objects on which dew forms must be good radiators and bad conductor, water vapour content of air are conditions required for frost formation. When the temperature of air falls below 0°C before the dew point is reached, the water vapour is directly converted into ice crystals and deposit on surface. At night, a blade of grass loses energy by emitting radiation while it gains energy by absorbing the energy emitted from surrounding objects. Under clear night time skies, objects near the ground emit more radiation than they receive from the sky, and so a blade of grass cools as its energy losses are greater than its energy gains. If the temperature of a grass blade gets cold enough and there is sufficient water vapor in the environment, frost will form on the grass.

Hoar frost is the deposition in the form of ice crystals having the shapes of scales, feather, needles, fern and it is white and opaque. Glazed frost is usually caused by rain which freezes on falling to the ground and is transparent.

Types of Frost

a) Based on Appearance

White frost is the deposition of condensed moisture in the form of white ice crystals upon a surface.

Black frost: When the temperature falls below freezing level, the water within the cells and intercellular spaces freezes and cause death of plant cells. The plant leaves look like burnt canopy.

b) Based on Process of Cooling of Air

Advection frost is formed due to advected cool air mass.

Radiation frost is formed due to radiational cooling of air.

8.3.3 Fog

The temperature of air decreases to such an extent that water droplets remain suspended in air but do not deposit over surface is called fog. It occurs in early morning hours. Extremely small water droplets suspended in the atmosphere and reducing the horizontal visibility is fog.

Classification of Fog

Thick fog: restricts visibility upto 45 meters.

Moderate fog: restricts visibility upto 450 meters.

Thin fog: restricts visibility upto 900 meters.

8.3.4 Mist

If the drops in a drizzle completely evaporates before reaching the ground is known as mist. Mist is less dense fog. It disappears with rising sun. The suspended water droplets restrict visibility between 1000-2000 meters. Relative humidity is at least 75 per cent mist disappear with rising Sun.

8.3.5 Cloud

It is mass of tiny droplets or ice crystals or both condensed on hygroscopic nuclei and suspending in atmosphere.

8.3.6 Smog

It is the combined effect of smoke, dust and fog droplets reduces the visibility.

8.3.7 Rime

When wet fog having super cooled droplets immediately freeze on striking objects and ice formed on windward side.

Chapter 9

Precipitation

9.1 Introduction

Precipitation is a condensed moisture in liquid or solid forms falling on the earth surface. It is earthward falling of water droplets or ice crystals which are formed because of rapid condensation and are too large to remain suspended in the atmosphere. Rainfall is the quantity of water falling on unit area of earth surface in specified time. The lines connecting various locations, having an equal amount of precipitation is known as Isohyets.

9.2 Mechanism or Processes of Rain Formation

There are two mechanisms by which raindrop can be formed;

a) Bergeron Mechanism

In cold cloud, ice particles and water drops are formed due to very low temperature (-15 to -25°C). These ice particles are grown by deposition of water vapour (sublimation) developing into hexagonal shaped ice particles. These ice crystals on collision form snow pellets and melt into water droplets, when falling on ground through a warm atmosphere. It is discovered by Bergeron in 1933. When ice crystals collide with super cooled droplets and on contact with ice crystal it causes droplets to freeze, this process is called accretion or riming and the riming matter is called grapple. When ice crystal masses grouped collide with other solid ice particles masses and ice crystals are broken up into fragments or tiny ice particles masses is called splintering. When such ice crystal become so large that cannot remain suspended in the atmosphere and come down.

b) Collision and Coalescence Mechanism

In hot clouds, fine water droplets exist instead of ice particles. These fine water droplets collide, coalescence and grow into the large size and fall on earth as rain drop. Coalescence process of rain making was discovered by E. G. Boven.

9.3 Types of Precipitation

Precipitation results from cloud due to adiabatic cooling of rising air. Precipitation is classified on the basis of conditions which cause rising of moist air masses into three types;

a) Conventional Precipitation

The precipitation, which results due to thermal convection of warm and moist air masses is called convectional precipitation. Due to heating, the air near the ground becomes hot and light and start upward movement is known as convection. As air moves upward, it cools at the dry adiabatic lapse rate, saturated adiabatic lapse rate, relative humidity increase to 100 per cent and dew point is reached where the condensation begin. First cloud is formed and then the further condensation results into precipitation. Rain, snow and hails are associated with this type of precipitation under favorable conditions it occurs in tropical and temperate zones. This type of precipitation mainly occurs in equatorial low-pressure belt.

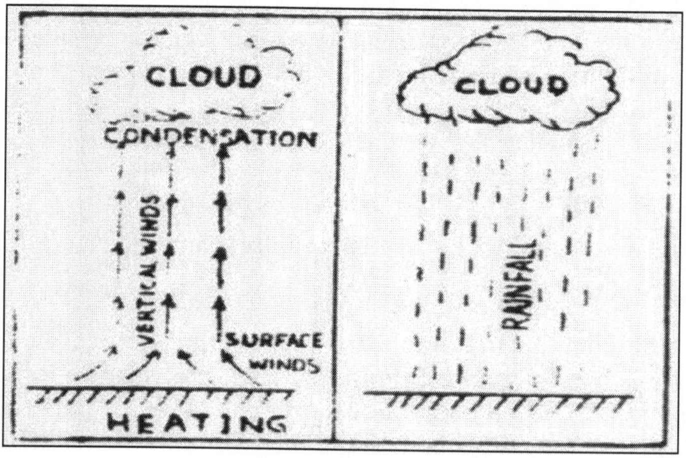

Figure 9.1. Conventional Precipitation.

b) Orographic Precipitation

The term orographic is derived from the Greek word '*oros*' means mountain. The precipitation resulted when air rises and cools due to topographic barrier is known as orographic precipitation.

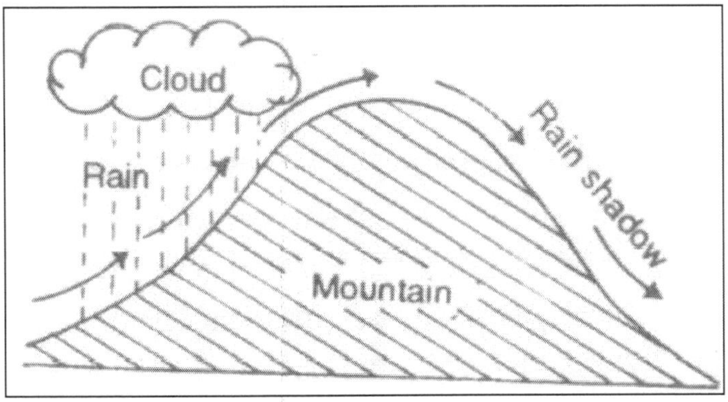

Figure 9.2. Orographic Precipitation.

When mountain comes in the path of air, the air has to pass that obstacle, while rising along the slope its cools, condenses and forms clouds which gives precipitation. This is known as orographic precipitation. This type of rainfall occurs in foot hills of Himalaya ranges but not in Aravali hill ranges, as they are parallel to the monsoon winds. This type of precipitation is on the windward side of mountains and exists dry area on the leeward sides of mountain ranges which is called as rain shadow.

c) Cyclonic Precipitation

Figure 9.3. Cyclonic Precipitation.

It is caused by lifting associated with the horizontal convergence of in flowing atmosphere into an area of low pressure. There are two kinds of cyclonic precipitation; i. Non frontal precipitation involves only convergence and lifting and

ii. Frontal precipitation results when one air mass is lifted over another. Cyclonic precipitation occurs when deep and extensive air masses of different characteristics converge, rise and adiabatic cooling takes place. In tropical regions where opposing air currents with different temperatures converge, vertical rising takes places and usually accompanied by convection. This convection causes condensation and precipitation.

In temperate regions mostly frontal precipitation occurs. The warm and moist air gradually rises over cool and dry air along the front. The convectional activity along the front cause's condensation and precipitation. This type of precipitation is called frontal precipitation.

9.4 Front

Front is a boundary which separates two opposite air masses having contrasting characteristics in terms of air temperature, humidity, pressure, wind direction, density. Transitional zone between two converging air masses is called frontal zone which represents zone of discontinuity. The process associated with formation of new fronts is called frontogenesis. The region having convergence of contrasting air mass is called region of frontogenesis. The destruction of fronts is called frontolysis.

9.4.1 Conditions Required for Frontogenesis

- Presence of two opposing air masses having contrasting properties.
- Convergence of two contrasting air masses leads to formation of fronts.
- Two opposing air masses must have contrasted temperature *i.e.,* one air mass should be cold, dry and dense, while the other should be warm, moist and light.

9.4.2 Classification of Fronts

a) Warm Front

It is frontal surface along which warm and light air becomes active and aggressive and rises slowly over cold and denser air causing extensive areas of cloudiness and precipitation. Warm fronts move at a speed of 15-50 km/hr.

b) Cold Front

It is frontal surface along which cold air becomes active and aggressive and in which the warm air territory being denser remains at the ground. Cold fronts move at a speed of 30-80 km/hr.

c) Occluded Front

It is formed when cold front overtakes warm front and warm air is completely displaced from the ground surface. The precipitation pattern is a combination of both warm and cold frontal distribution. Occluded fronts move at a speed of from 8-50 km/hr.

d) Stationary Front

It is formed when two contrasting air masses converge in such a way that they become parallel to each other and there is no ascent of air.

9.5 Air Masses

An air mass is an extensive portion of the atmosphere which is approximately homogeneous in its horizontal distribution of temperature, humidity, and lapse rate. Regions where air masses originate are referred to as source regions. The area is not sufficiently homogeneous to act as source regions. These portions of the earth serve as transition zones.

Bergeron introduced the concept of air masses in forecasting practice and has laid down the basic principles of air-mass analysis. This classification separates air masses into four basic types:

- ✰ Equatorial air masses {E), typical of the source region for equatorial air.
- ✰ Tropical air masses (T), typical of the source region for tropical air.
- ✰ Polar air masses (P), typical of the source region for polar air.
- ✰ Arctic air masses {A), typical of the source region for arctic air.

The thermodynamic feature of Bergeron's classification consists of two main types:

- ✰ Cold air masses (k), indicating air that is colder than the underlying surface.
- ✰ Warm air masses {w), indicating air that is warmer than the underlying surface.

The thermodynamic classification indicates stability or instability conditions in the lower layers of the air mass, and the geographical classification indicates the source region of the air mass. The temperature and moisture content of a parcel of air can be uniquely expressed in terms of the pseudo-wet-bulb potential temperature; or the equivalent potential temperature, it has become customary to identify the various air masses by their equivalent potential or their pseudo-wet-bulb potential temperatures.

9.6 Forms of Precipitation

The form of precipitation which falls to the earth surface depends upon the; dew point temperature, conditions of atmosphere through which particles pass, types of clouds and processes of precipitation.

- ✰ **Rain:** It is a precipitation in the form of liquid water droplets. The diameter of liquid water drops more than 0.5 mm. When these minutes water droplets in clouds combine and form large drops that becomes so large that they cannot remain suspended in the air and they fall down as rain.

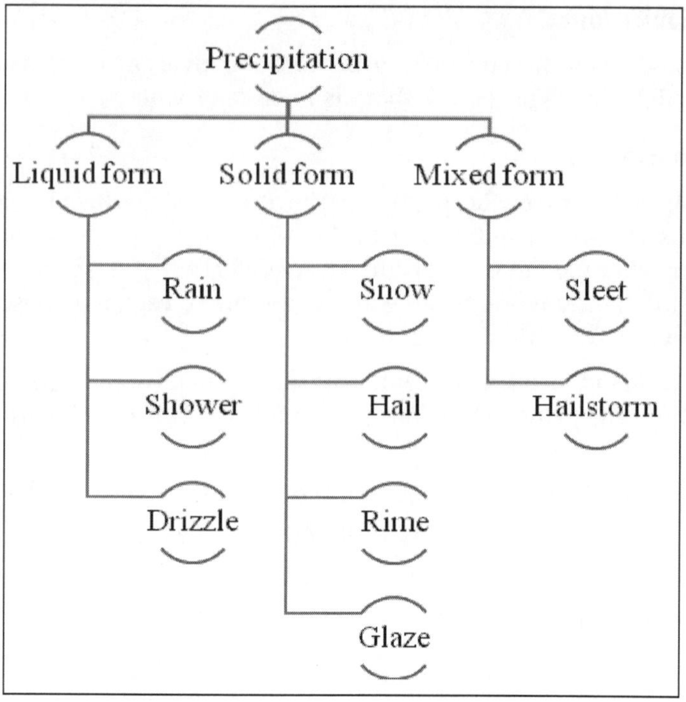

- ☆ **Drizzle:** It is a precipitation in the form of fine water drops having diameters less than 0.5 mm. If the drops in a drizzle completely evaporated before reaching the ground, the condition is referred to as mist.
- ☆ **Shower:** Precipitation lasting for a short time with relatively clear intervals is called shower. This occurs from the passing clouds.
- ☆ **Snow:** It is a precipitation in the solid form of small or large ice crystals. It occurs only when the condensing medium has a temperature below freezing temperature. Snowflakes are formed in high clouds. Snow is measured with snow gauge.
- ☆ **Hail:** It is a precipitation of condensed atmospheric moisture in the form of small bolls or pieces of ice having diameters more than 5 mm.
- ☆ **Hailstorm:** Rainfall associated with hailstones known as hailstorm.
- ☆ **Sleet:** It is a precipitation of the mixture of rain and snow.
- ☆ **Rime:** White, opaque, granular deposit of ice crystals formed on object that are at a temperature below the freezing point.
- ☆ **Graupel:** It is a small particles of snow with a fragile crust of ice.
- ☆ **Glaze**: It is a smooth, transparent homogeneous ice coating occurring when freezing rain/drizzle hits a surface.

9.7 Measurement of Rainfall

9.7.1 Instruments Used to Measure Rainfall

Rainfall is collected and measured by means of rain gauges. The rainfall measurement is taken as the vertical depth of water that would accumulate on a level surface if the rainfall remains where it fell.

Types of Rain Gauge

- Non-Recording Rain Gauge
- FRP (Fibre Reinforced Plastic) type rain gauge
- Ordinary rain gauge
- Recording Rain Gauge
- Tipping Bucket Type rain gauge
- Weighing Bucket Type rain gauge
- Natural Siphon Type rain gauge

A) Non Recording or Ordinary Type Rain Gauge

a) USWB Type Rain Gauge

This is the simplest and economical non-recording type rain gauge. It consists of a cylindrical receiver of 8-inch diameter provided with a funnel shaped bottom. The rain caught by the receiver is transmitted to the cylindrical measuring tube 20 inches long and one tenth of cross-sectional area of receiver. The depth of rain fall is thus then magnified ten times and measured to an accuracy of $1/10^{th}$ of an inch. Rainwater entering the gauge from the top of the rim of the funnel is lead to the receiver. The rainwater thus collected is measured with the help of measuring cylinder.

b) Fibre Glass Reinforced Polyester Rain Gauge

The non-recording rain gauge used in India is the Symon's gauge. The rain gauge has a circular collecting area of 12.7 cm diameter connected to a funnel. The collector rim is set in a horizontal plane at a height of 30.5 cm above the ground level. The base and collector are made of fiber reinforced plastic (FRP); hence this rain gauge is also known as FRP rain gauge. The rainfall catch is discharged into a receiving vessel by means of the funnel. The funnel as well as the receiving vessel are housed in a metallic container. Water collected in the receiving vessel is measured by a graduated cylinder having an accuracy up to 0.1 mm. The measuring cylinder is calibrated according to the area of the funnel. The area of measuring cylinder is $1/10^{th}$ area of the collector. For uniformity, the rainfall is measured every day at 8.30 AM (IST) and is recorded as the rainfall of that day. The receiving bottle normally does not hold more than 10 cm of rain and as such in the case of heavy rainfall the measurements must be done more frequently and entered. However, the last reading must be taken at 8.30 AM and the sum of the previous readings in the past 24 hr entered as total of that day. Proper care, maintenance and inspection of rain gauges, especially during dry season. Ordinary rain gauge can also be used to measure snowfall.

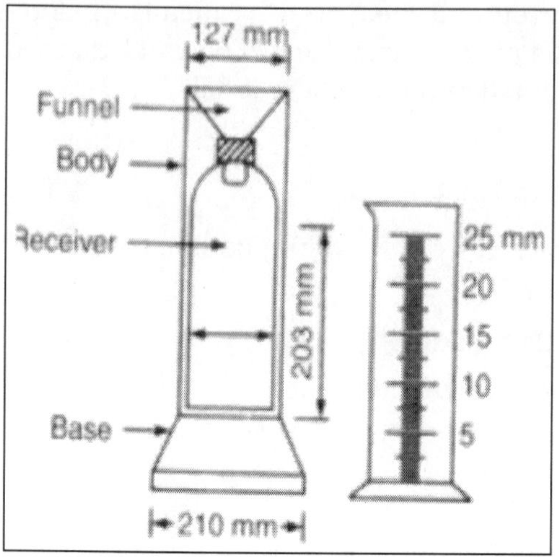

Figure 9.4. Fibre Glass Reinforced Polyester Rain Gauge

B) Automatic/Self-Recording Type Rain Gauge

a) Electric Tipping Bucket Type Rain Gauge

This is an electric transmitting rain gauge. These rain gauges are generally attached to the automatic weather stations. It is a 30 cm sized circular rain gauge adopted for use by USWB. It has 30 cm diameter sharp edged receiver and at the end of the receiver is provided a funnel.

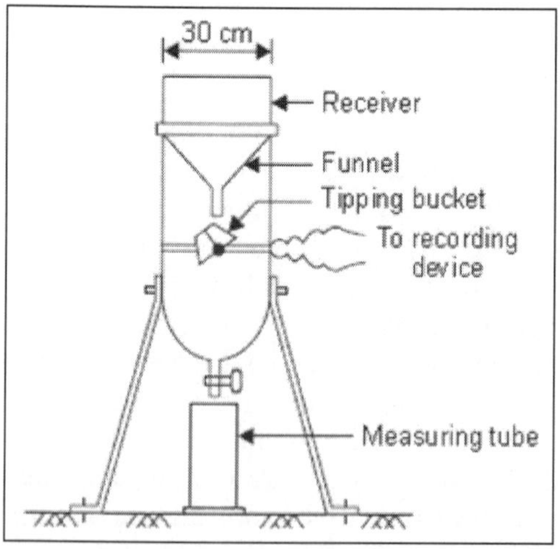

Figure 9.5. Electric Tipping Bucket Type Rain Gauge.

Pair of buckets are pivoted under this funnel in such a manner that when one bucket receives 0.25 mm of precipitation (rainfall), it tips discharging its rainfall into the container, bringing the other bucket under the funnel. Tipping of bucket completes an electric circuit causing the movement of pen to mark on clock driven receiving drum which carries a recorded sheet. These electric pulses generated are recorded at the control room far away from the rain gauge station. This instrument is further suited for digitalizing the output signal. This tipping mechanism also provides a method of recording on a moving chart or a counter, attached to the tape recorder in the weather station.

b) *Recording Weighing Type Rain Gauge*

It is most common self-recording rain gauge. It consists of a receiver bucket supported by a spring or lever balance or some other weighing mechanism. The weight of the water or snow during precipitation moves a platform downward. An 8-inch diameter cylinder is fixed on a platform. The platform is arranged on the spring balance. The downward movement of the platform is feed to the pen and ink arrangement by lever. The pen makes the marking on the graph paper graduated on one axis in inches or centimeter of rain and time on another axis. This rain gauge can also be used to measure snow fall. The instrument directly records the rate of fall in inches of rainfall per unit time for any desired interval of time in addition to total fall since previous reading. Weighing bucket type rain gauge instrument gives a plot of the increased rainfall values against the elapsed time and the curve so formed is called the mass curve.

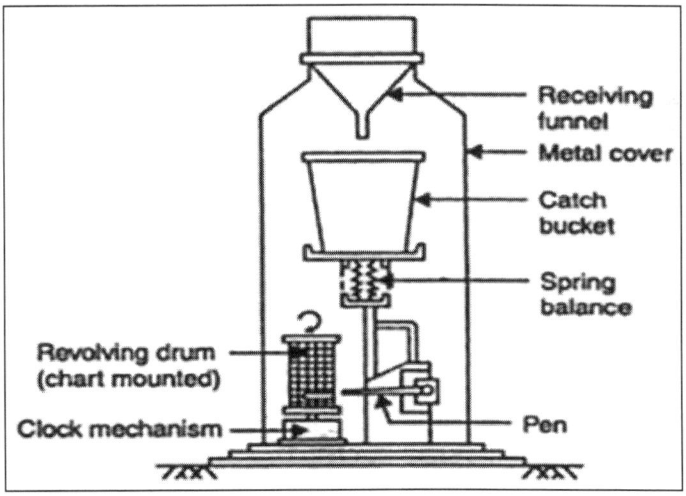

Figure 9.6. Recording Weighing Type Rain Gauge.

c) *Recording Float Type Rain Gauge*

The working of this type of rain gauge is similar to weighing bucket rain gauge. A funnel receives the water which is collected in a rectangular container. A float is provided at the bottom of container, and this float raises as the water level rises in the container. Its movement being recorded by a pen moving on a recording drum actuated by a clock work. When water rises, this float reaches to the top floating in water, then syphon comes into operation and releases the water outwards through the connecting pipe, thus all water in box is drained out. This rain gauge is adopted as the standard recording rain gauge in India and the curve drawn using this data is known as mass curve of rain fall. Every morning the graph paper needs to be replaced. Siphoning occurs after every 10 mm of rainfall.

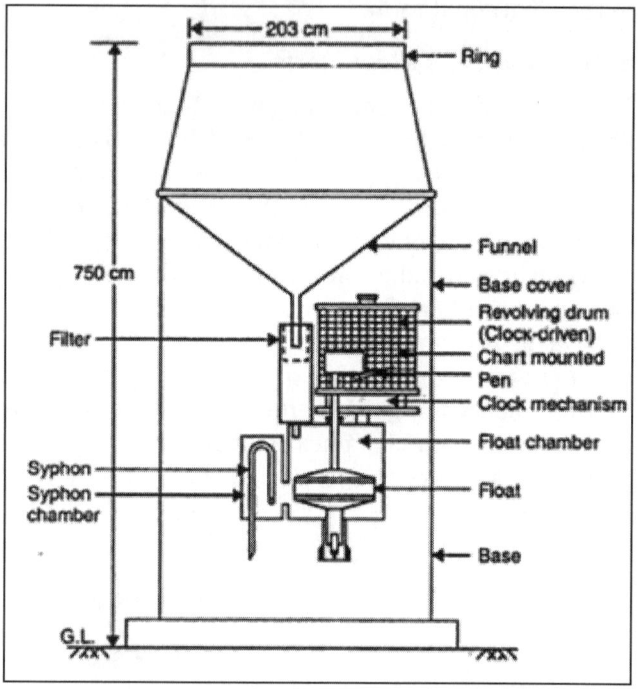

Figure 9.7. Recording Float Type Rain Gauge.

9.7.2 Important Points Related to the Rainfall

Seasonal Rainfall Distribution

Sl.No.	*Category*	*Range of Rainfall*
1	Large Excess	Percentage departure of actual rainfall from normal rainfall is +60 per cent or more
2	Excess	Percentage departure of actual rainfall from normal rainfall between +20 per cent to +59 per cent
3	Normal	Percentage departure of actual rainfall from normal rainfall between +19 and +19 per cent

Sl.No.	*Category*	*Range of Rainfall*
4	Deficient	Percentage departure of actual rainfall from normal rainfall between -20 and -59 per cent
5	Scanty	Percentage departure of actual rainfall from normal rainfall between -60 and -99 per cent
6	No Rain	Percentage departure of actual rainfall from normal rainfall is -100 per cent

Rainfall Distribution on Macro Scale

Sl.No.	*Category*	*Range of Rainfall*
1	Normal	Percentage departure of actual rainfall from normal rainfall is within -10 and +10 per cent
2	Below Normal	Percentage departure of actual rainfall from normal rainfall is <10 per cent of the long period average
3	Above Normal	Percentage departure of actual rainfall from normal rainfall is >10 per cent of the long period average
4	Deficient Year	When the rainfall deficiency is more than 10 per cent and 20-40 per cent area of the country is under drought condition.
5	Large Deficient Year	When the rainfall deficiency is more than 10 per cent and 40 per cent or more area of the country is under drought condition.
6	Drought Year	When the rainfall deficiency is 20 per cent
7	Severe Drought Year	When the rainfall deficiency is 40 per cent

Spatial Distribution of Rainfall

Sl.No.	*Category*	*Range of Rainfall*
1	Widespread	75 per cent or more number of stations of a region reporting at least 2.5 mm rainfall
2	Fairly Widespread	51-74 per cent number of stations of a region reporting at least 2.5 mm rainfall
3	Scattered	26-50 per cent number of stations of a region reporting at least 2.5 mm rainfall
4	Isolated	25 per cent or less number of stations of a region reporting at least 2.5 mm rainfall
5	Dry	No station of a region reported rainfall

Intensity of Rainfall

Sl.No.	*Category*	*Range of Rainfall*
1	No rain	Rainfall amount realized in a day is 0.0 mm
2	Trace	Rainfall amount realized in a day is between 0.01 and 0.04 mm
3	Very light rain	Rainfall amount realized in a day is between 0.1 and 2.4 mm
4	Rainy Day	Rainfall amount realized in a day is more than or equal to 2.5 mm
5	Light rain	Rainfall amount realized in a day is between 2.5 and 15.5 mm
6	Moderate Rain	Rainfall amount realized in a day is between 15.6 and 64.4 mm
8	Heavy Rain	Rainfall amount realized in a day is between 64.5 and 115.5 mm
9	Very Heavy Rain	Rainfall amount realized in a day is between 115.6 and 2404.4 mm
10	Extremely Heavy Rain	Rainfall amount realized in a day is more than or equal to 204.5 mm

9.8 Artificial Precipitation

The precipitation received by artificially saturation of a special type of cloud is known as artificial precipitation. or it is the precipitation by artificial seeding of clouds with hygroscopic nuclei. Cloud seeding is a type of weather modification that aims to change the amount or type of precipitation that falls from clouds by dispersing substances into the air that serve as cloud condensation or ice nuclei. The chemicals used for cloud seeding include silver iodide, potassium iodide, dry ice (solid carbon dioxide) and liquid propane. The cloud seeding is of two types based on the temperature of clouds;

a) Seeding of Cold Clouds

The temperature of clouds below the freezing point is called as cold clouds. Fine powder of silver iodide is seeded in the cloud which acts as nuclei and produces ice particles in clouds. Silver iodide smoke is also introduced in clouds by aero planes or explosive rockets and balloons.

b) Seeding of Warm Clouds

The temperature of clouds above the freezing point is called as warm clouds. Fine salt particles or dry ice is seeded in the clouds which act as hygroscopic nuclei and produce rain drops in clouds.

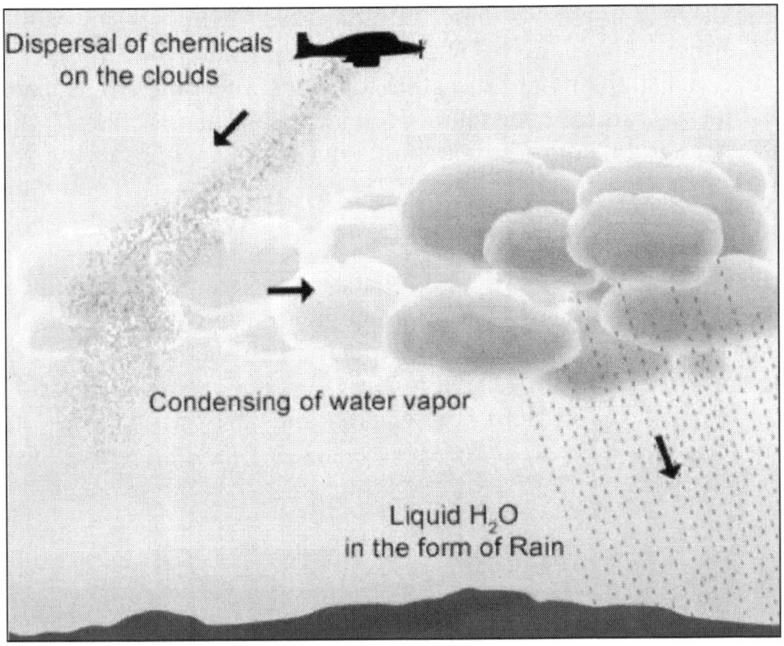

Figure 9.8. Cloud Seeding.

9.9 Monsoon

Monsoon is a wind pattern that reverses direction with the seasons. The word monsoon comes from the Arabic word "*mawsim*", the Malayan word *'monsin'*, the Dutch word *'monssoen'*, the Portuguese word *'monçao'* which means 'season'.

9.9.1 Monsoon Systems

Monsoon is caused by seasonal cycle of temperature over land as compared to the adjacent oceans. The hot air over the land tends to rise and creating area of low pressure. Contrary to this the high-pressure center exists over sea surface. The rainfall is caused by moist ocean air which is drawn to the land. In winter, the land cools quickly but the ocean surface remains warmer and the center of low and high pressure shifted and hence the monsoon circulation also changes its direction. There are two types of monsoon system over India; south west Monsoon (June - September) and north east Monsoon (October - December).

a) South West Monsoon

During the hot and dry season (month of April and May), when temperature rises rapidly and pressure over land decreases, the warm and moist air from the adjacent ocean start blowing towards the low-pressure center. By the end of May or first week of June when low pressure center is fully developed and the pressure gradient is steeple, the trade winds from the southern hemisphere are drawn towards the thermal low in northern latitudes (India). India receives 80 per cent of rainfall during summer monsoon. The southern trade wind when crosses the equator deflected to their right direction, then the S-E trade winds become S-W winds. The normal rainfall over this season in India is about 88 cm (Jun-Sept). It reaches first in Kerala; hence Kerala is called as gateway of India monsoon. The SW monsoon has two branches;

i. *Arabian Sea Branches*

It strikes the western ghats of India at almost right angle. The windward slope receives heavy orographic precipitation, however the westerlies current from Arabian sea continues its journey across Indian peninsula. But amount of rainfall continues to decrease on leeward side with distance from sea coast. These air currents ultimately unite with Bay of Bengal branch in northern India.

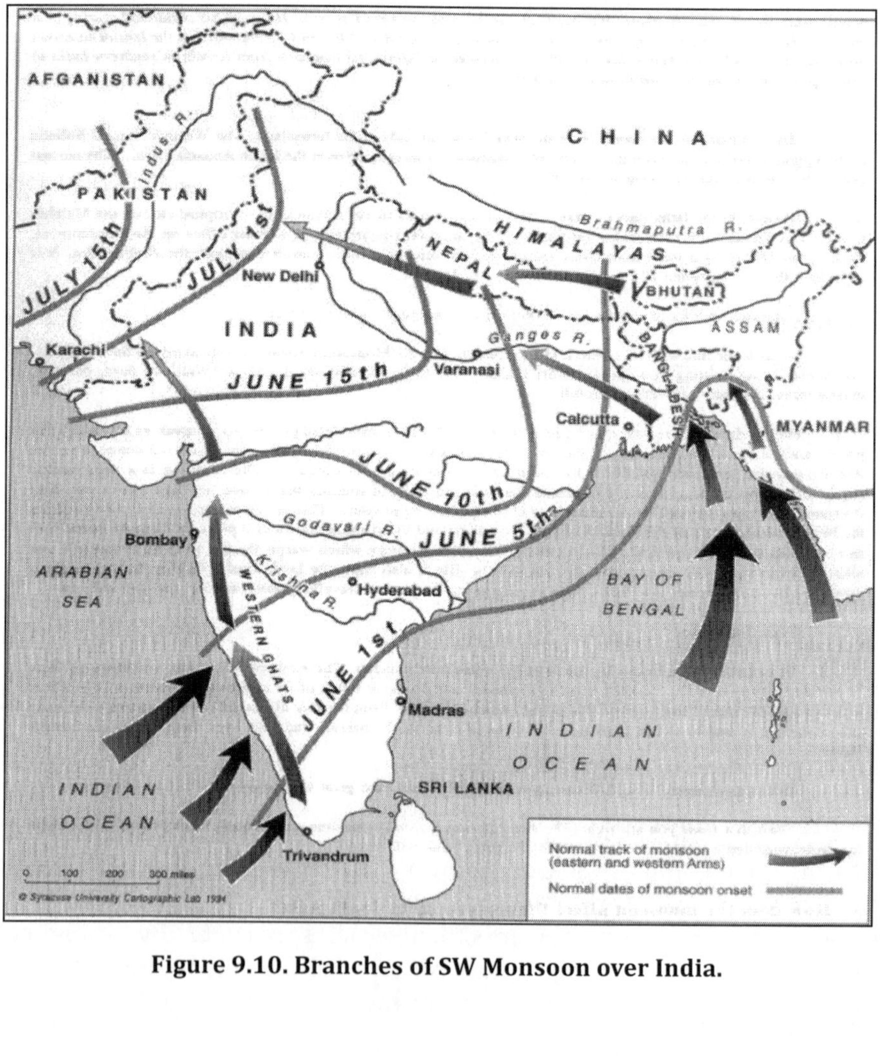

Figure 9.10. Branches of SW Monsoon over India.

ii. Bay of Bengal Branches

One current of Bay of Bengal branch moves towards Assam and cause heavy rainfall in Mawsynram (near Cherrapunji) situated on the Khasi hills. One current of BOB branch recurved and advances up to the Gangetic plane of Punjab. It is to be noted that rainfall decreases from east to west direction. The main reason of decrease in rainfall in west ward direction is because of increase in distance from the source of moisture.

b) North East Monsoon

Secondary high-pressure system develops over Kashmir and the Punjab in winter. The high-pressure area controls the prevailing wind direction over the rest of the subcontinent. Contrary to the pressure condition over land, there are low pressure centers formed over the Indian Ocean, the Arabian Sea, and northern part of Australia. In the cool season, therefore, there is pressure gradient from land to sea as a result of which winds begin to move from land to sea. These are the Northeast or winter monsoons of northern hemisphere. The southern part of Indian Peninsula receives rainfall from Northeast monsoon currents. These currents while traveling over the Bay of Bengal pick up moisture from warm ocean surface. The amount of winter rainfall on the eastern side of the peninsula is much heavier than that on the other side. Tamil Nadu receives maximum rainfall in this season. It is also known as retreating monsoon. The normal rainfall during this monsoon is around 12 cm.

9.9.2 Importance of Monsoon in Indian Agriculture

- Monsoon is the lifeline of Indian Agriculture. A good monsoon brings economic prosperity for the whole country and boosts the Indian economy as agriculture contributes around 16 percent of the total GDP.
- A normal rainfall is important for adequate agricultural production in order to maintain food prices.
- A large number of Indian farmers are still dependent on the monsoon rains for various agricultural activities.
- In the absence of a good monsoon, the entire nation may be destabilized because of food inflation.
- Several dams, reservoirs, rivers and canals in India are rain-fed and depend upon the monsoon rains.
- Economic activities like forestry and fishing are also influenced directly by the amount of rainfall.
- Monsoon rain gives us respite from the high and intolerable summer heat and brings back life to the living earth.
- The Monsoon rains in India also replenish reservoirs and groundwater that helps in improving irrigation and also boosts hydropower production.
- The amount of rainfall from monsoon effects on choice of crops, cropping system, farming system in particular area.

9.10 Hydrological Cycle

The continuous movement of water on, above and below the surface of the earth is called as hydrological cycle or water cycle. Continuous circulation of water between hydrosphere, atmosphere and lithosphere is known as hydrological cycle.

The main source of moisture to the atmosphere is through the process of evaporation from oceans, river, lakes, well *etc.* The water vapour condenses and form clouds and falls through precipitation in the form of rain, drizzle, snow, ice pellets and hails. River and streams flow back to the ocean where water vapour again has its origin. Hydrological cycle is a never-ending process or it is continuous process.

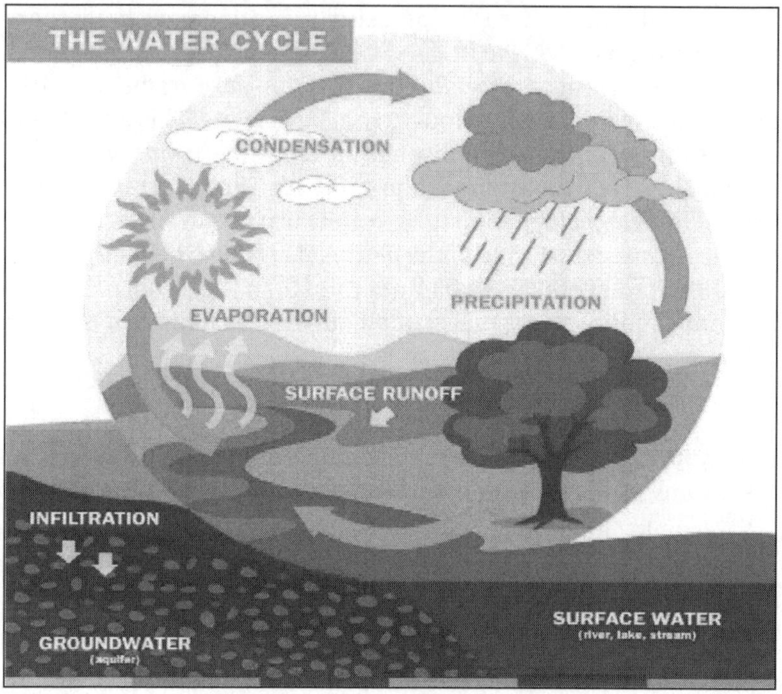

Figure 9.11. Hydrological Cycle.

9.10.1 Hydrological Balance Equation

Hydrological balance equation is a useful balance for parameters of water availability and water needs at a place expressed as

$$P + I = AET/PET + \Delta SM + RO + D$$

where,

P = Precipitation

I = Irrigation

AET = Actual evapotranspiration

PET = Potential evapotranspiration

SM = Change in soil moisture storage

RO = Runoff

D = Percolation losses.

a) Percolation and Infiltration

Percolation is the downward movement of the water through the soil due to force of gravity. Infiltration is the movement of water from the surface onto the soil. Infiltration rate is the soil characteristic determining the maximum rate at which water can enter the soil under specific conditions. According to the rate of entry of water from surface to the soil, infiltration rate is grouped into four categories;

- ☆ Very slow: Infiltration rate less than 0.25 cm/hr.
- ☆ Slow: Infiltration rate between 0.25 to 1.25 cm/hr.
- ☆ Moderate: Infiltration rate between 1.25 to 2.50 cm/hr.
- ☆ Rapid: Infiltration rate more than 2.50 cm/hr.

Factors Affecting Infiltration Rate

- ☆ **Compactness of soil surface:** A compact soil surface permits less infiltration, whereas, loose soil surface has more infiltration rate. Cultivation influences the infiltration rate by increasing he porosity of the surface soil and breaking up the surface seals.
- ☆ **Impact of rain drop:** When impact of raindrop is more than it causes scaling and closing of pores resulting reduction in infiltration rate.
- ☆ **Vegetative cover:** Soil surface with vegetative cover has more infiltration rate than bare soils. Infiltration rates on grassland is substantially higher than bare uncultivated land.
- ☆ **Soil moisture content:** In wet soils, infiltration rate is less and in dry soil, infiltration rate is more.
- ☆ **Soil temperature:** Warm soils absorb more water than cold soils.
- ☆ **Soil texture:** In coarse texture soils, infiltration is more as compared to heavy soils. In clay soils the cracking caused by drying also increases infiltration in the initial stage until the soil again swells and decreases infiltration.
- ☆ **Depth of soil:** Shallow soils permit less water to enter into the soil than too deep soils.
- ☆ **Organic matter content:** Higher the organic matter content in soil, higher the rate of infiltration.
- ☆ **Water depth over soil:** In surface irrigation increased depth increases initial infiltration slightly, but negligible effect after prolonged irrigation.

b) Runoff

It is the part of the water cycle that flows over land as surface water instead of being absorbed into groundwater or evaporating. There are three types of runoff;

a) Surface runoff: It is the portion of rainfall that enters streams, channels immediately after occurring rainfall. It is also called as prompt runoff as it takes very little to join the streams. b) Subsurface runoff: It is amount of rainfall, which first enters into soil and then flows laterally towards stream without joining water table. c) Base flow: It is the part of rainfall which after falling over the ground surface, percolate into soil and meet to water table and finally join to stream or ocean.

Factors Affecting Runoff

I) Climatic Factors

- **Types of precipitation:** Precipitation occurs in form of rainfall, start moving immediately in the form of overland runoff depending upon intensity and duration of precipitation. Precipitation occurs in form of snow or hails, the flow of water on the ground surface will not take place immediately, but after melting start runoff.
- **Rainfall duration:** Total runoff depends on the duration of rainfall. Longer the duration of rainfall more the runoff.
- **Amount of Rainfall**: If more rainfall hits the ground, more rainfall will turn into runoff. If a large amount of snow melts in a short time period, there will be a large amount of runoff.
- **Rainfall intensity:** Rainfall intensity influences both rate and volume of runoff. The runoff volume and runoff rate will be greater for an intense rainfall event than for less intense event.
- **Distribution of rainfall:** It influences both rate and volume of runoff. The maximum rate and volume of runoff occurs when the entire watershed contributes.
- **Direction of prevailing wind:** If the direction of wind towards the runoff flow in the drainage system of the catchments, then the resulting runoff will be more at short interval.

II) Physiological Factors

- **Shape and size of catchment:** More intense rainfall events are generally distributed over a relatively smaller area, *i.e.*, larger the area lower will be the intensity of rainfall. The peak normally decreases as the area of the basin increase. Larger basins give a more constant minimum flow than the smaller ones.

- ☆ **Vegetation**: Vegetation needs water to survive, and a plant's root system is designed to absorb water from the soil. There is less runoff in highly vegetated areas because the water is used by the plants instead of flowing off the surface of the ground.
- ☆ **Slope**: The steeper surface will faster the flow down the slope and a flat surface will allow the water time to absorb which reduces the runoff.
- ☆ **Permeability**: The ability of the ground surface to absorb water will affect how much surface runoff occurs. The less water the ground can absorb, the more runoff on the surface there will be. A surface with high absorption ability has high permeability, and a surface with low absorption ability has low permeability.
- ☆ **Storage characteristics:** Presence of artificial storage such as dams, weirs *etc.* and natural storage such as lakes and ponds tend to reduce the peak flow.
- ☆ **Orientation of watershed**: Windward side of mountains get more rainfall than leeward side due to orientation effects.
- ☆ **Soil moisture**: More soil moisture causes less infiltration results in increasing runoff. Dry soil absorbed more water results in less runoff.
- ☆ **Soil type**: Light soil involves rapid absorption of water, thus it has less runoff potential. Heavy textured soils causing less infiltration and more runoff.

c) Evaporation

It is the process in which water is lost to atmosphere from water body in the form of water vapour. It is the process during which a liquid changes into a gas. It is diffusive process. Evaporation rate is the volume evaporated per unit area in a unit time.

Factors affecting Evaporation

- ☆ **Temperature:** As the temperature increases, the rate of evaporation also increases. Temperature and rate of evaporation are proportional to each other. With an increase of temperature, the kinetic energy of water molecules increases and surface tension decreases.
- ☆ **Surface area:** As the surface area increases, the rate of evaporation increases. The surface area and rate of evaporation are proportional to each other.
- ☆ **Atmospheric humidity:** The rate of evaporation decreases with an increase in humidity. Humidity and rate of evaporation are inversely proportional to each other. When the air has low relative humidity, the

evaporation will be more, when air has high relative humidity over water evaporation rate will be low.

- **Wind velocity:** Increase in wind speed results in increased in evaporation. Wind speed and rate of evaporation are proportional to each other. The process of evaporation takes place continuously when there is a supply of energy to provide latent heat of evaporation.
- **Air pressure:** If air pressure is high on the surface of a body of water, then the water will not evaporate easily.

The pressure pushing down on the water makes it difficult for water to escape into the atmosphere as vapor.

- **Composition of Water:** The dissolved salts and other impurities affect the rate of the evaporation. The evaporation is inversely preoperational to the salinity of water. The rate of evaporation from the surface of the sea is less than that of fresh water in rivers.
- **Vegetative cover:** More the vegetative cover on the surface of the soil less the evaporation rate as compared to less vegetative cover soil.

d) Transpiration

It is loss of water in the form of water vapour through plants. There are three kinds of transpiration; stomatal, cuticular and lenticular transpiration.

Types of Transpiration

i) **Stomatal Transpiration:** Water vapour diffuses out through stomata present in soft aerial part of plant is known as Stomatal transpiration. Stomatal transpiration constitutes about 80-90 per cent of the total water lost during transpiration. Stomata are minute pores present in the epidermis of aerial parts of the plant body that remain guarded by two guard cells.

ii) **Lenticular Transpiration:** It is the process of water loss in the form of water vapor through the internal tissues of plant body through the internal tissues of the plant body about 0.5 per cent of total water lost during transpiration. Lenticels are minute pores present in the periderm of the stem of plants and consist of loosely arranged complementary cells.

iii) **Cuticular Transpiration:** It is the process of loss of water in the form of water vapor through the internal tissues of the plant body through the cuticular surface of the plant body. The cuticle is a waxy coating present on the epidermis of aerial parts of the plant body. It constitutes about 3-10 per cent of the total water lost during transpiration. This type of transpiration depends upon the thickness of the cuticle and presence or absence of wax coating on the surface of the leaves. The thinner the cuticle, the higher the rate of transpiration. Xerophytic plants generally have very thick cuticle and wax coating on the leaves and stem in order to check cuticular transpiration. In sciophytes, cuticular transpiration contributes about 50 per cent of the total water lost during transpiration.

Factors Affecting Transpiration

- ☆ **Atmospheric humidity:** The rate of transpiration is inversely proportional to relative humidity. More the relative humidity less is the transpiration rate.
- ☆ **Light:** The direct effect of light is on the opening of stomata. The indirect effect of light is that the increasing light intensity raise the temperature of leaf cell. The stomata open during the day and close in the dark. Presence of light is directly proportional to the rate of transpiration.
- ☆ **Air temperature:** A high temperature lowers the relative humidity and opens the stomata even in darkness. As a result, the rate of transpiration increases.
- ☆ **Wind:** If wind is still, the transpiration rate is low. This is because the water vapour accumulates around the transpiring organs and reduce the diffusion pressure deficit of the wind. If the wind is moving, the saturated wind around the leaves is removed and transpiration rate increases.
- ☆ **Wind velocity:** Fast moving wind bring fresh and dry masses of air in contact with leaf surfaces, higher the wind speed higher the transpiration.
- ☆ **Plant height:** The rate of transpiration of a tall crop will be more than cut or clipped crops.
- ☆ **Leaf size:** The rate of transpiration is reduced as the size of leaf reduced.
- ☆ **Availability of water**: The transpiration rate is directly proportional to the absorption of water by the roots from the soil. A decrease in water absorption causes the closure of stomata and wilting, thereby reducing the rate of transpiration.

e) Evapotranspiration

It is the quantity of water transpired by plants and the moisture evaporated from the surface of the soil. Potential evapotranspiration is a measure of the ability of the atmosphere to remove water from the surface through the processesof evaporation and transpiration assuming no control on water supply. Actual evapotranspiration is the quantity of water that is actually removed from a surface due to the processes of Evapotranspiration.

Measurement of Evapotranspiration

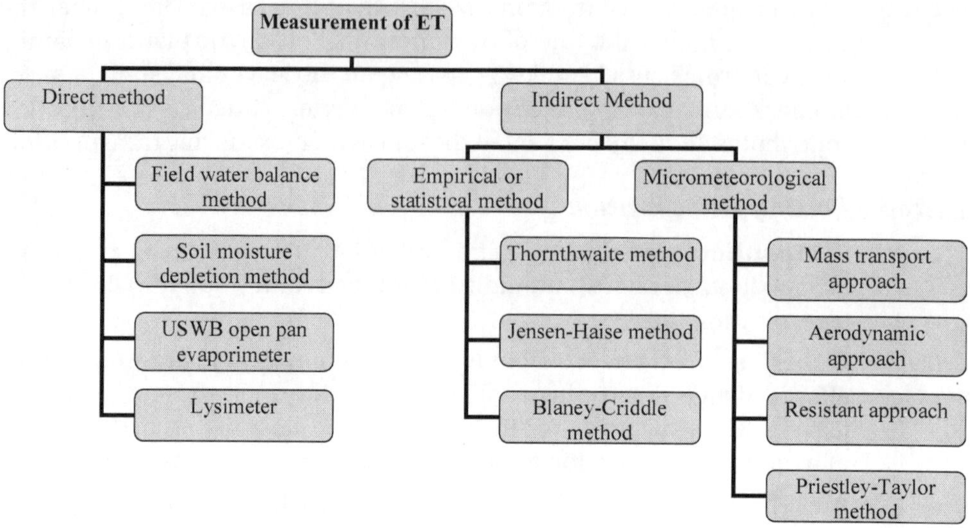

USWB Class a Pan Evaporimeter

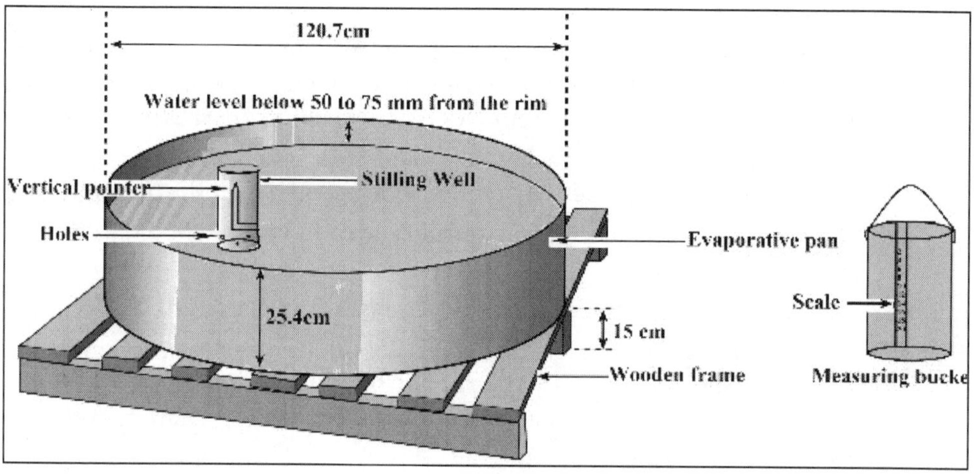

Figure 9.12. USWB Class a Pan Evaporimeter.

The physical process by which any liquid escape from the surface into the atmosphere in gaseous state is called evaporation. The standard United States Weather Bureau (USWB) Class-A open pan evaporimeter is most commonly used to measure evaporation from free water surface. The evaporimeter is placed on a wooden stand. It consists of a 120.7 cm diameter and 25.4 cm deep pan made of 20-gauge galvanized iron sheet with a stilling well of size 10 cm diameter and 30 cm height placed within the tank. A vertical pointer is provided in the stilling well to show the level of water maintained in the pan. Its purpose is to isolate a small portion of the water surface in the tank so that it is not disturbed by waves produced by wind. Three small holes are located at the bottom of the well to permit the flow of water in and out of the well. The pan is painted white and is placed on a wooden frame so that air may circulate beneath the pan. A measuring cylinder is used to measure the evaporation rate. The cross-sectional area of the measuring cylinder is exactly 1/100 of the area of evaporation pan. The scale from 0 to 20 cm is engraved inside it along the height and the graduation runs from top to bottom in ascending order. One full cylinder (200 mm) of water raises 2 mm height in the pan. It is measured by unit Millimeter and observed on 08:30 hrs, 02:00 hrs and 05:30 hrs.

Measuring Procedure

1. No Rainfall

When there is no rainfall, measured quantity of water is added to the evaporation pan up to the tip of the pointer using measuring bucket. The evaporation is equal to the amount of water added to the pan.

Evaporation = Total volume of water added by cylinder/100

2. Rainfall with pointer above water level

When there is rainfall, but the tip of the fixed-point gauge remains above the water level, measured quantity of water is added by measuring bucket till the water level reaches tip of the pointer in the stilling well.

Evaporation = Water added/100 + Rainfall during the period (mm)

3. Rainfall with pointer below water level

When rainfall is more than evaporation and the water level inside the stilling well is above the tip of the pointer, in this case water is removed from reservoir with the help of measuring bucket until the level of water comes to the tip of the pointer.

Evaporation = Rainfall during the period (mm) – (Water removed/100)

4. Overflow

When there is heavy rainfall, over flow will take place from the pan, evaporation cannot be recorded.

Precautions

- If the rainfall is heavy, water must be removed from the tank with measuring cylinder. Difference between the actual rainfall of the previous day and water removed from the tank gives the evaporation rate.
- If there is very heavy rainfall, the tank overflows and evaporation value cannot be obtained. The message overflow or excess rainfall is written in the weather report.

Sunken Screen Pan Evaporimeter

Sharma and Dastane (1968) developed the sunken screen pan evaporimeter that provides evaporation values closer to crop ET values.

The evaporimeter consists of a 60 cm diameter and 45 cm deep pan made of 20-gauge galvanized iron sheet and stilling well of 15 cm diameter and 45 cm depth attached to the pan with a tube. The pan and the well are painted white and screened at the top with 6/20-mesh wire net. The stilling well has a pointer inside at the center. The pan with the well is buried in the soil with 10 cm edge over the soil surface. The fall of water level in the well during 24-hour period is taken as the measure of the evaporation rate per day. Further, evaporation rate may be computed from the measured quantity of water added daily to bring the water level to the pointer tip level after each measurement. Tip of the pointer should be at level with the soil surface while installing the evaporimeter. The pan is located in the field with no obstruction to wind movement over the pan.

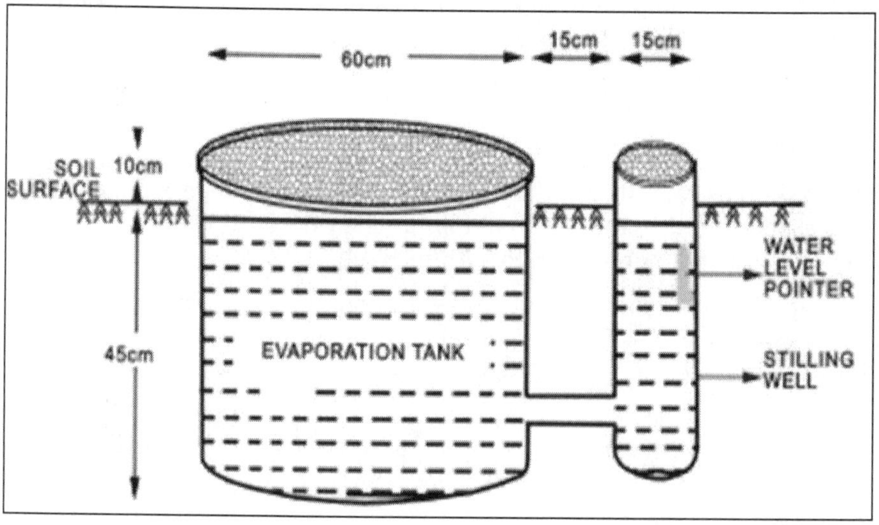

Figure 9.13. Sunken Screen Pan Evaporimeter.

Difference between USWB class a pan and Sunken screen pan evaporimeter

Points	*USWB Class a Pan Evaporimeter*	*Sunken Screen Pans Evaporimeter*
Place of instruments	Above the ground surface	Below the ground
Diameter of pan	122 cm	60 cm
Height of pan	25 cm	45 cm
Place of stilling well	Inside the pan	Outside the pan
Crop coefficient	0.5 t 1.3	0.95 to 1.05
Rate of evaporation measured	At free surface	In the cropped field

Lysimeter

There are two types of lysimeters: 1) Volumetric lysimeter which measure ET by volume *e.g.* floating lysimeter, drum typre rice lysimeter. 2) Gravimetric or weighing lysimeters which measure ET by weight.

Gravimetric or Weighing Type Lysimeters

It consists of the lysimeter tank (1.3 x 1.3 x 0.9 m) containing soil in which the plants are grown is mounted on the platform (120 x 120 cm) of weighing balance (2000 kg). The steel tank carries a perforated plate at a depth of 75 cm so as to form a hallow chamber. A tube is inserted through the perforated plate into the bottom of hallow chamber through this tube the percolated water is removed.

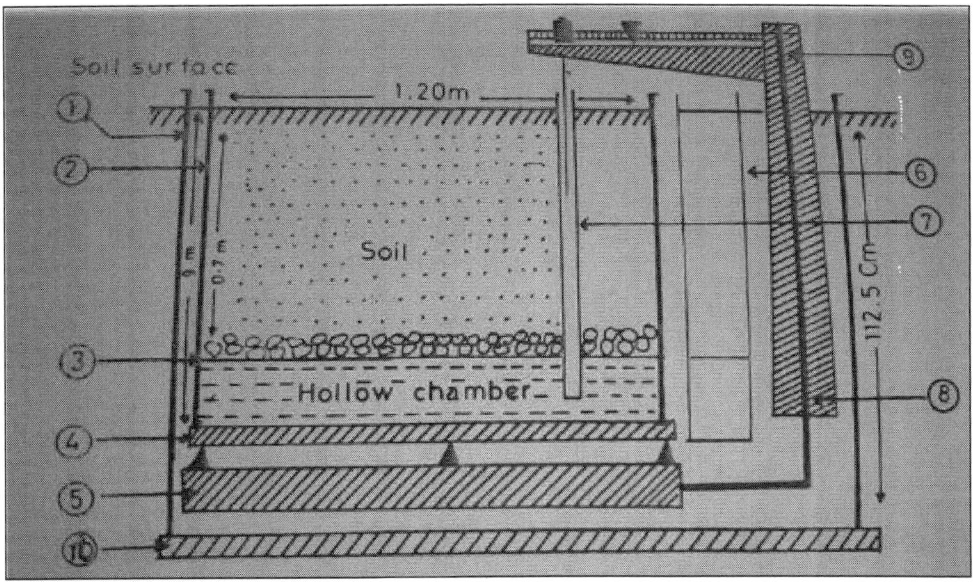

Figure 9.14. Schematic Diagram of a Gravimetric or Weighing Type Lysimeter.

(1. Retaining tank, 2. Lysimeter tank, 3. Perforated plate, 4. Platform, 5. Balance, 6. Dummy tank, 7. Pipe to remove percolated water, 8. Balance, 9. Balance, 10. Foundation).

The retaining tank (140 x 140 x 112.5 cm) protect them from the water seepage from the adjoining field and A dummy tank is prevent the overheating of lysimeter tank side which is near the pillar of weighing machine. It works on the principle of an isolated block of cropped soil representative of the field is weighed first to determine quantity of water added by precipitation or irrigation and then after loss of water by ET is again weighted. The difference of weight in kg in the two observations multiple by conversion factors 0.6 gives the ET in mm.

Soil Water Balance

$$ET = I + P - RO - DP + CR +/- \Delta SF +/- \Delta SW$$

Evapotranspiration can be determined by measuring the various components of the soil water balance. The method consists of assessing the incoming and outgoing water flux into the crop root zone over some period. Irrigation and rainfall add water to the crop root zone. Water from irrigation (I) and rainfall (P) might be lost by surface runoff (RO) and deep percolation (DP) that will eventually recharge the water table. Water might also be transported upward by capillary rise (CR) from a shallow water table towards the root zone or even transferred horizontally by subsurface flow in (SF_{in}) or out (SF_{out}) of the root zone. Soil evaporation and crop transpiration depletes water from root zone.

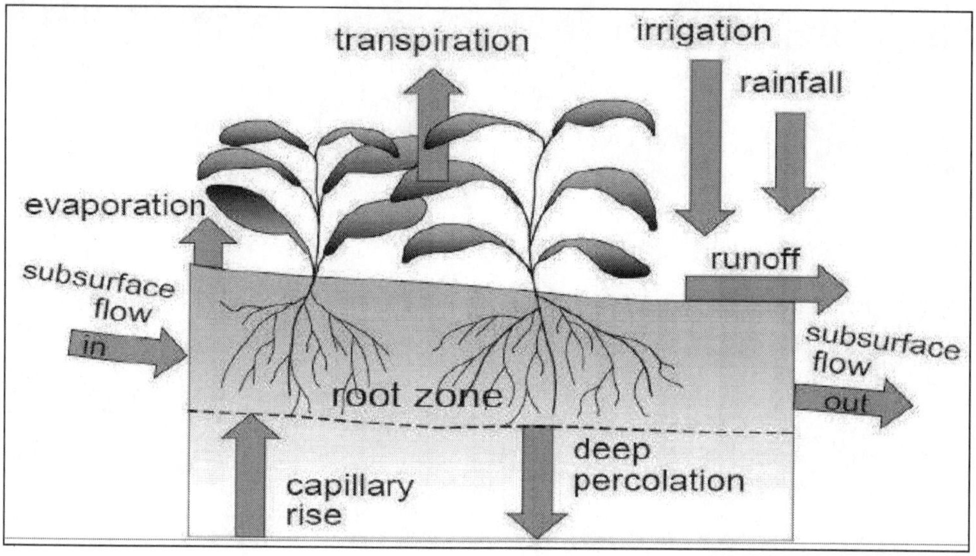

Figure 9.15. Soil Water Balance at Root Zone.

Chapter 10

Cloud Formation and Classification

10.1 Definitions

Clouds are visible aggregate of liquid water droplets and ice crystals suspended in air. The branch of meteorology which studies clouds is called nephology.

10.2 Cloud Formation

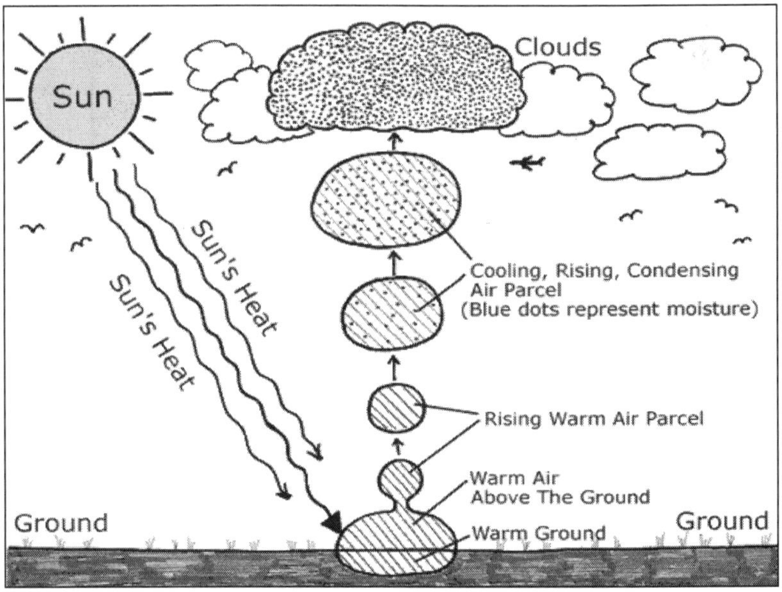

Figure 10.1. Process of Cloud Formation.

It is a form of condensation. Presence on conventional currents in lower layer of troposphere causes formation of clouds. Sun heats the water bodies causes evaporation of water. Warm air rises upper part of troposphere then cooling of air occurs with dry adiabatic lapse rate and saturated adiabatic lapse rate. The water vapour in moist air gets condensed on condensation nuclei. These condensed masses of particles form a cloud. Due to buoyancy effect of air, this mass remains floating in upper layers of air. (Buoyancy effect - All air particles in the atmosphere are drawn by the downward force of gravity. But the pressure in the air creates an upward force working opposite gravity pull. Air density builds to whatever level balance the force of gravity because at this point gravity is not strong enough to pull down a greater number of particles.

10.3 Cloud Classification

In 1803, Luke Howard is first person to classify the clouds on the basis of their structure and named the different clouds with Latin words: Cirrus, cumulus, Nimbus and Stratus.

The International Cloud Atlas published by World Meteorological Organization describes the ten main genera or types.

10.3.1 Classification of Clouds on the Basis of their Appearance and Process of Formation

a) Genera

These are main characteristic forms of clouds. The ten cloud genera are cirrus, cirrocumulus, cirrostratus, altocumulus, altostratus, nimbostratus, stratocumulus, stratus, cumulus and cumulonimbus.

b) Species

The peculiarities in shape and differences in internal structure of clouds. The fourteen cloud species are fibratus, uncinus, spissatus, castellanus, floccus, stratiform, nebulosus, lenticularis, fractus, humilis, mediocris, congestus, calvus, and capillatus.

c) Varieties

The special characteristics of arrangement and transparency of clouds. The nine cloud varieties are intortus, vertebratus, undulatus, radiatus, lacunosus, duplicatus, translucidus, perlucidus, and opacus.

d) Supplementary Features and Accessory Clouds

Appended and associated minor cloud forms. The nine supplementary features and accessory clouds are incus, mamma, virga, praecipitatio, arcus, tuba, pileus, velum, and pannus.

e) Mother Clouds

The origin of clouds if formed from other clouds.

10.3.2 Classification of Clouds on the Basis of their Composition

a) **Water clouds** are composed entirely of water droplets. *e.g.* altocumulus, stratus, nimbostratus, stratocumulus, and cumulus.

b) **Ice crystal clouds** are composed of ice. *e.g.* cirrus and cirrostratus.

c) **Mixed clouds** are a combination of water and ice crystals. *e.g.* cumulonimbus, cirrocumulus and altostratus.

10.3.3 Classification of Clouds on the Basis of their Height or Altitude

Family and Height	*Name and Abbreviations*	*Composition*	*Appearance and Description*
Family A (7-12 km)	Cirrus (Ci)	Ice crystals lock or tuft of hair	Delicate detached **fibrous feathery**, silky appearance, sun shine without shadow, does not produce precipitation, white in colour, layered, tufty, patchy in shape.
	Cirrocumulus (Cc)	Mixedcirro (curl of hair), cumulus (heaped)	Arranged as ripples in the sky, look like scaly skin of a fish called **mackerel sky**, doesn't produce precipitation, sun shines without shadow.
	Cirrostratus (Cs)	Ice crystals stratus (layered)	**White veil**, milky appearance, transparent, produce halo, doesn't produce precipitation, sun shine without shadow.
Family B (3-7 km)	Altocumulus Ac	Mixed alto (middle)	Arranged in regular pattern of waves or parallel bands or rounded masses, **sheep clouds or wool pack clouds,** doesn't produce precipitation.
	Altostratus As	Mixed	**Fibrous veil** of grey or bluish grey, produce corona phenomena around sun and moon, clouds cover entire sky, do not produce precipitation but occasionally light showers.

Family and Height	*Name and Abbreviations*	*Composition*	*Appearance and Description*
Family C (0-3 km)	Stratus	Water	Cover the sky in a blanket of white or grey sheet or layer, produce corona phenomena, light precipitation or drizzle, clouds are broken then it called as fractostratus.
	Stratocumulus	Water	Long parallel rolls, pushed together or broken masses which look soft and white to dark grey, formed due to flattening of cumulus clouds.
	Nimbostratus	Mixed nimbus (rainy clouds)	Dark, grey, featureless layer of clouds, produce continuous rain, it never accompanied by lightening, thunder or hail. Streaks of water or snow falling from these clouds but not reaching the ground are called "Virga".
Family D Clouds with vertical development	Cumulus	Water	Clouds are fluff, cauliflower in shaped, occasionally rain or snow occurs, look like wool pack or **pieces of floating cotton,** clouds dark below due to shadow.
	Cumulonimbus	Mixed **(King of clouds)**	Only clouds that produce hail, thunder and lightning, towering anvil top, violent up and draft, cloud electrification occurs due to many collisions between charged water droplets, groupel and ice crystals, chief precipitation makers.

10.4 Measurement of Clouds

A Ceilometer is a device that uses a laser or other light source to determine the height of a cloud and also used to measure the aerosol concentration within the atmosphere. The Laser Ceilometer measures cloud height and thickness. A laser pulse is emitted into the atmosphere and backscatter analyzed. Using the speed of light, the altitude of each cloud base and top is determined. An Okta is a unit of measurement of cloud used to describe the amount of cloud cover at any given location. Cloud cover is measured on an eight-point scale, 0 Oktas (clear sky), 1 Octa (1/8 of the sky covered in cloud) and so on up to 8 Oktas (completely overcast). A Ceiling balloon is used to determine the height of the base of clouds above ground level during day light hours.

Chapter 11

Weather Forecasting

11.1 Introduction

Weather forecasting is any advance information about the probable weather in future obtained by evaluating the present and past meteorological conditions of the atmosphere. Prediction of weather in advance is called as weather forecasting. Agrometeorological forecast is the advance information regarding weather related to agriculture and farmers is known as.

11.2 Types of Weather Forecasting

Weather forecasting on the basis of their validity period or time scale classified as;

a) Now Casting

It denotes very short range *i.e.* few hours to 24 hours. *e.g.* Forecasting at the time football, cricket match during the day.

b) Short Range Forecasting (SRF)

It is valid for 3 days or 72 hours and are issued twice a day based on synoptic conditions. It includes cloud spread, rainfall distribution, heavy rainfall warning, minimum and maximum temperature, heat and cold waves, low pressure area, cyclone warning, hail, thunder, snow, frost and wind speed. It is useful in preventive measures against adverse weather.

c) Medium Range Forecasting (MRF)

It is valid for 3 to 10 days period. The forecast includes cloud amount, rainfall, temperature, average wind speed and wind direction. A National Center of Medium Range Weather Forecasting (NCMRWF) was established in 1988 to develop atmospheric model for MRF. Agricultural operations like sowing, planting, spraying, dusting, irrigation scheduling, storing, fertilizer application, transportation, protection from frost, hails *etc.* can be forecasted.

d) Long Range Forecasting (LRF)

It is valid for more than 10 days. The IMD issuing LRF since the year 1988 onwards on total monsoon rainfall of the country by 25th May. These forecasts can be used for predicting trends in food grains production of India before beginning of the season.

11.3 Utility of Weather Forecasting

- Helps for suitable planning of farm
- Help to decide sowing time
- Help for farm operation;
    i. To irrigate the crop or not
    ii. To apply fertilizer
    iii. To start harvesting.
- Helps in transportation and storage of grains
- Helps for management of cultural operations
- Helps to protect livestock

11.4 Methods of Weather Forecasting

a) Synoptic Method

The data of meteorological elements such as atmospheric pressure, temperature, humidity, rainfall, wind, clouds, visibility, weather phenomena and waves are recorded at surface and various heights above the earth surface at various meteorological stations situated throughout the country and sent to main center through electromagnetic signals. A large volume of meteorological observations from surface and aloft obtained from a dense network of stations are depicted in pictorial and coded form on the weather maps or charts by drawing isopleths (lines joining places with equal numerical value of the weather element). The inferences on expected movement of weather system are drawn using the previous and present synoptic charts. This method of forecast is mainly used for short range weather forecasting by IMD.

b) Analogue Method

This method is defined as any particular type of system if it is present in the past also, it is assumed that the present system is likely to behave in the same way as the previous one.

c) Persistence Method

It assumes that the conditions at the time of the forecast will not change and yesterday's weather may continue to behave in same manner for today also. Often summarized as "Tomorrow equals today".

d) Climatological Method

This method involves historical weather data over long periods of time (years) to predict conditions on a given date. The weather statistics accumulated over many years has been averaged to make the weather forecast.

e) Statistical Methods

This method is based on statistical approaches such as regression and auto regressive integrated moving average techniques. The regression equations are used to predict weather parameters. Knowledge of correlation coefficient will also help to access effect of one parameter over the other. This method is used for long range forecasting of Indian monsoon rainfall. It is introduced operationally by IMD in 1988 as 16 Parameters Power Regression Model.

11.5 Tools of Weather Forecasting

Pilot balloons, radiosonde, radar and satellite, synoptic charts, crop weather calendar and crop weather diagram are the tools for weather forecasting.

a) Pilot Balloons

A small balloon is inflated with a gas hydrogen or helium and released for measurement of wind direction and speed at different height in the atmosphere using a theodolite. The theodolite is an optical telescope which measures horizontal and vertical angles of the ascending balloons at known intervals of time. A radio-theodolites are used to track the balloon using radio signals to overcome balloon lost and no longer traced problems. Meteorographs were attached to the balloon for measure temperature, humidity and pressure at different heights in the atmosphere. The meteorograph is a combination of barograph, thermograph and hygrograph.

b) Radiosonde

The radiosonde is a device with sensitive sensors which transmits observations by radio signals to a ground receiver as the balloon ascends. It is carried aloft by means of a free balloon. The instrument measures temperature (+40 to -90°C), humidity (15 to 100 per cent) and pressure (1040 to 40 kpa).

c) Radar and Satellite

They are the two space tools to sense meteorological parameters using remote sensing techniques. The radar (radio detection and ranging) transmits electromagnetic pulses of a given characteristics and also receive them back. It is used for detecting tropical cyclones and monitoring their course also used for detecting cloud amount and precipitation. The Indian satellite (INSAT) is a domestic satellite system used in communication, television, radio broadcasting and meteorology. It gives information on cloud spread and temperature of the earth surface. This information is used in short range and medium range weather forecasting.

d) Synoptic Chart

It is a chart or map on which various weather data and analysis are presented that describes the atmosphere over a large area at a given time.

e) Crop Weather Calendar

It is detailed information for each important crop on their dates of sowing, dates of commencement and duration of major cultural operations, important periods in their life cycle and their most probable weather requirements have been presented in a pictorial form. It is a tool to a weather forecaster for providing efficient weather services. The crop weather calendars consist of three parts. At the bottom is the typical life history of the crop, from the sowing window to the period of maturity, in the form of a diagram. The middle of the calendar shows the weekly normal weather as far as such conditions could be determined from long term averages. The uppermost portion of the calendar indicates the weather conditions the nature of the weather warnings or weather abnormalities to take precautionary measures. Top part is divided into different section according to dry spell, length, high wind, heavy rainfall and cloudy weather *etc.*

f) Crop Weather Diagram

It provides season wise information on crop husbandry (tillage to harvest), actual and normal weather months or weeks, pest and disease incidence. It is a tool to assess the crop conditions and it's yield in relation to weather. It helps in understanding favorable weather conditions that are responsible for better crop yield through agronomic manipulation and predict crop yield quantity based on weather conditions. Crop weather diagram as a tool to assess the crop condition and its yield in relation to weather, base for improving the quality of crop weather calendars and to a weather forecaster, for providing efficient weather service.

11.6 History of Weather Forecasting

1636	A British scientist Halley published his treatise on the Indian summer monsoon, which he attributed to a seasonal reversal of winds due to the differential heating of the Asian land mass and the Indian Ocean.
1785	The British East India Company established station at Calcutta for studying the weather and climate of India.
1784	The Asiatic Society of Bengal founded at Calcutta to promote scientific studies in meteorology in India.
1796	The British East India Company established station at Madras (now Chennai) for studying the weather and climate of India.

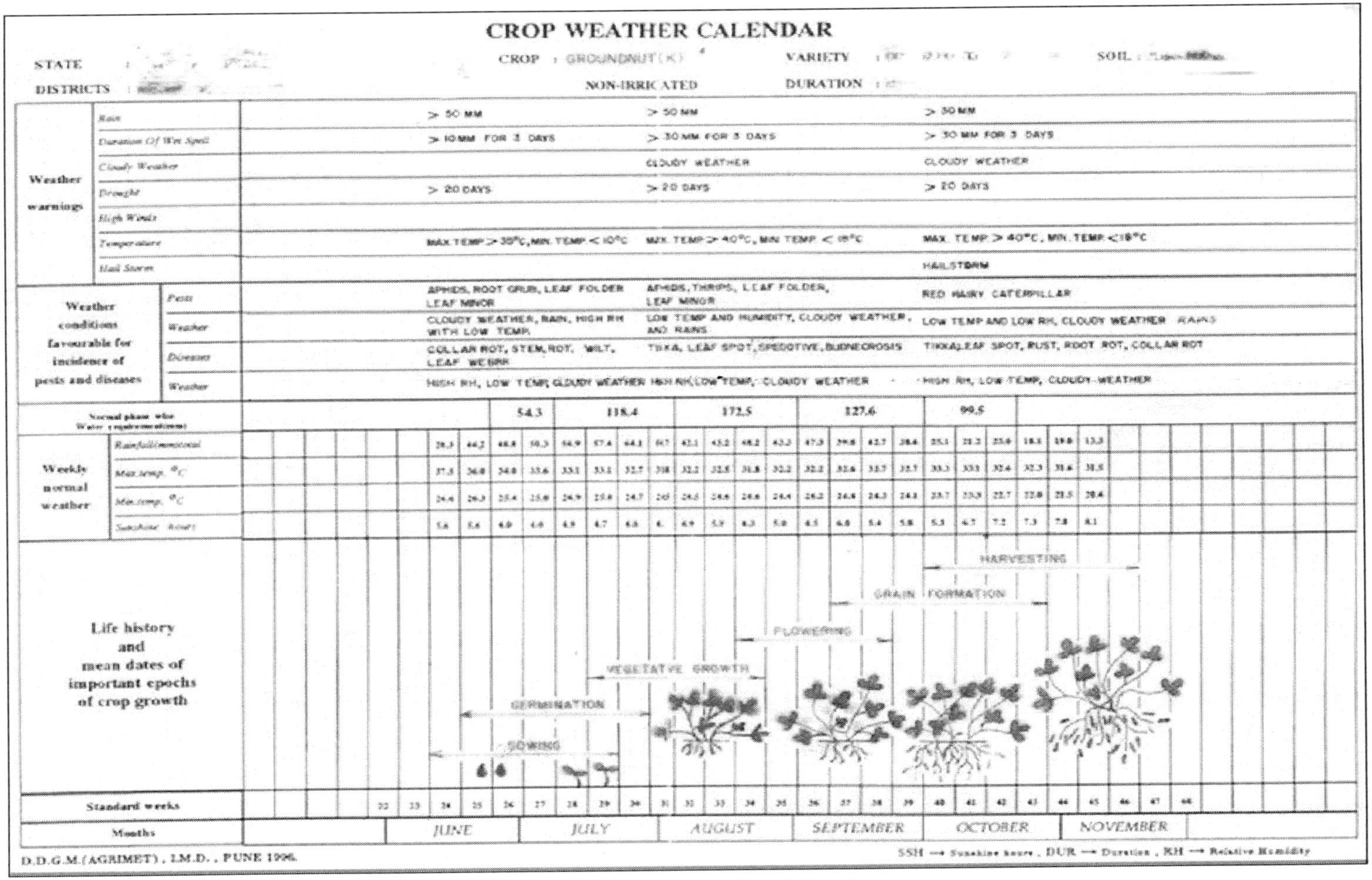

Figure 11.1. Crop Weather Calendar.

1804	The Asiatic Society of Bengal founded at Bombay (now Mumbai) to promote scientific studies in meteorology in India.
1835-1855	Captain Harry Piddington at Calcutta published 40 papers during in the Journal of the Asiatic Society dealing with tropical storms and coined the word "cyclone", meaning the coil of a snake.
1842	Captain Harry Piddington published his monumental work on the "Laws of the Storms".
1875	The Government of India established the India Meteorological Department, bringing all meteorological work in the country under a central authority.
1875	**Mr. H. F. Blanford** was appointed Meteorological Reporter to the Government of India.
1886	IMD issued first time seasonal forecast of rainfall.
1889	**Sir John Eliot** who was appointed the first Director General of Observatories in at Calcutta headquarters.
1905	Upper air measurement started in India using tracking balloons with theodolite.
1909	Walker developed multiple regression models for forecasting monsoon rainfall over India.
1913	Routine observation from pilot ballon was stared.
1927	Establishment of the Imperial Council of Agricultural Research which is now the Indian Council of Agricultural Research at New Delhi.
1922	English scientist Lewis Fry Richardson published "Weather Prediction by Numerical Process".
1924	Walker established the link between the Indian monsoon with global meteorological phenomenon.
1928	K. R. Ramanathan published the diagram of the distribution of upper atmospheric distribution over the globe.
1932	The division of Agricultural Meteorology was started in order to provide direct services to the farming community of the country under the umbrella of India Meteorological Department at Pune to minimize the impact of adverse weather on crops and to make use of favorable weather to boost agricultural production. It became a permanent department of IMD in 1943. The prime object of Agricultural Meteorology Division, IMD, Pune is to minimize the impact of adverse weather on crops and to make use of crop-weather relationships to boost agricultural production.
1947	IMD became a member of the World Meteorological Organization on 27th April.

1945	All India Radio started broadcasting a farmer's weather bulletin.
1948	IMD and ICAR had launched a crop weather scheme for compilation of phenological data and weather
1962	Indian Institute of Tropical Meteorology was established at Pune. The main organization responsible for meteorological observations, weather forecasts, and detecting earthquakes in India.
1974	The first KVK was established in Pondicherry.
1975	The IMD is initiated AAS in collaboration with State Agricultural Department.
1976	Agro-meteorological advisories were first initiated to provide state level forecast.
1983	AICRP on Agro-meteorology was started by ICAR under CRIDA.
1988	National Centre for Medium Range Weather Forecasting (NCMRWF) was established to provide Medium Range Weather Forecasts through deterministic Methods and to offer Agro-Advisory Service to the farmers.
1988	Government of India started a project "Crop Acreage and Production Estimation (CAPE)" to collect the statistics of Agricultural Output.
2007	The Integrated Agro-meteorological Advisory Service (IAAS), was introduced. It involves a wide range of partners, including the India Meteorological Department, the National Centre for Medium Range Weather Forecasting, and the Indian Council of Agricultural Research.
2008	District Agro-meteorological Advisory Service was launched, with the aim of providing relevant weather information and management advisories at a district scale across the country.
2016	The IMD launched System of Aerosol Monitoring and Research (SAMAR) to study the concentration of Black carbon, radiative properties of aerosols, environmental visibility and their climatological impacts.

11.7 Study of Agro-Met Advisory Services

Agro-meteorological Advisories are the single window solutions for all the queries related to fluctuating weather conditions and management of crops according to that. Agro-advisory services are the farm decisions taken in response to past, current and future weather change. It includes agronomical, pest and disease, water and input management. Basic considerations to prepare weather based agro-advisories are weather sensitive crops, their weather sensitive stages and weather sensitive farm operations. A Krishi Vigyan Kendra is an agricultural extension center in India. KVKs have implemented technologies to provide farmers information, such as weather advisories or market pricing, through radio and mobile phones.

The Indian Council of agricultural Research and India Meteorological Department jointly planned to provide Agro Met Advisory Services in local language across the country. ICAR and IMD have jointly planned to augment the AAS network to sub-district (block level) level by setting up of District Agro-Met Units (DAMUs) in 530 districts in the premises of KVKs to provide agro-met services to farmers. Agro advisory bulletin include three parts; Part 1: Summary of weather of the past week; Part 2: Weather forecast of coming 5 days from IMD. Forecast given to acquaint the farmer with expected weather on cloud amount, rainfall average, wind speed, wind direction, maximum, minimum temperature, relative humidity *etc.*; Part 3: Advisories for farmers. The agro advisory bulletins are prepared every Tuesday and Friday and it is communicated to farmers same day through Institute and KVK websites, WhatsApp messages and also through Facebook. The meteorological services provide weather data and 5-day forecasts. Specialists from ICAR, state departments of agriculture and the universities translate these into agricultural advisories to alert farmers to weather-related events that are likely to affect their farming operations, such as strong winds, low temperatures or periods of humid weather, which can increase the risk of disease outbreaks. They also provide advice on what actions farmers should take. Field units at the agricultural universities relay the advisories to farmers in local languages using a variety of channels, including SMS messages on mobile phones, local radio and newspapers, and face-to-face advisory and extension services. The IAAS also generates national-level and state-level advisory bulletins used by government planners and the agro-input supply industry.

Objectives of AAS

- To improve the existing district level Agromet Advisory Services for delivering crop and location specific AAS to farmers at block level advisories.
- To design optimum observatory network for issuance of village level advisories to implement crop weather insurance.
- To establish DAMU as nodal center for catering to needs of agriculture services.
- To provide customized advisory bulletins through last mile connectivity to farmers with personalized agro met advisory services.
- To extend the weather based advisory service to the allied areas like livestock, grazing of farm feed *etc.*
- To establish appropriate dissemination and support system for weather-based crop insurance in the country.

11.7.1 Gramin Krishi Mausam Sewa

Under GKMS project, India Meteorological Department, Ministry of Earth Science in collaboration with State Agricultural Universities/Indian Council of

Agricultural Research *etc.* is issuing crop and location specific weather based agro advisories for the benefit of farming community on every Tuesday and Friday and occurrence of extreme weather. The primary objective is preparation of weather-based advisories for management practices in agriculture after due consultation with subject matter specialists of concerned disciplines and disseminate the same to the end users up to village level using all possible modes of communication.

Under this district level weather forecast for next 5-days in respect of Rainfall, maximum temperature, minimum temperature, wind speed, wind direction, relative humidity and clouds, weekly cumulative rainfall forecast and Crop specific advisories are provided to farmers. GKMS of IMD has been successful in providing crop specific advisories to farmers through different print/visual/Radio/IT based media including short message service (SMS) and Interactive Voice Response Service (IVRS) facilitating appropriate field level actions. Based on the weather forecast, Agromet Advisories are prepared in collaboration with 130 Agromet Field Units (AMFUs) and communicated to the farming community. AAS rendered by IMD is a step to weather-based crop and livestock management strategies and operations dedicated to enhancing crop production besides reducing crop damage and loss due to extreme weather events. The GKMS of IMD has been successful in providing the crop specific agro meteorological advisories in vernacular languages to about 22 million farmers in the country.

11.7.2 Forecasting Agricultural Output using Space, Agro-Meteorological and Land Based Observations

FASAL was designed to collect Monsoon data through remote sensing, economic data and monitoring of crops when growing. The programme is sponsored by Ministry of Agriculture. A team of ISRO/Department of Space, State Remote Sensing Applications Centers, State Agricultural Universities and many other institutions are working on this project. Under the operational component of FASAL programme of Ministry of Agriculture and Farmers Welfare, MNCFC regularly generates crop forecasts at District/State/National level for 9 major crops of the country (such as Jute, *Kharif* Rice, Sugarcane, Cotton, Rapeseed and Mustard, *Rabi* Sorghum, Wheat, *Rabi* Pulses and *Rabi* Rice), using the procedures developed by Space Applications Centre, ISRO. Both optical and microwave Remote sensing data is used for crop acreage estimation, crop condition assessment and production forecasting. Collaboration with State Agriculture Department and State Remote Sensing Centers to carry out the data analysis, crop cutting experiment and ground truth data collection.

National/State/District level crop production forecasts were generated using multi-date AWiFS NDVI product/LISS-III/Sentinel 2/Landsat-8 and multi date Sentinel 1A SAR data for acreage estimation and estimation of crop yield, using multiple approaches; Semi-physical remote sensing based model, Remote Sensing index based yield models, Correlation weighted district level Agro-met model of IMD, Crop simulation model by IMD and Data obtained from crop cutting experiments.

11.7.3 Agro Meteorological Field Units

The IMD has set up in the country a network of about 130 Agro Meteorological Field Units which are multidisciplinary units responsible for preparation and dissemination of district and sub-district agro-met advisories. These AMFUs are located at State Agricultural Universities, ICAR centers and other institutions. Each AMFU utilizes the relevant output products including weather data from conventional/automatic weather station (AWS) provided by IMD and ICAR to generate specific advisories for agricultural management for the respective districts of Agro-climatic zones identified under the area of its jurisdiction and disseminate the same to the farming community.

11.7.4 District Agro-met Units (DAMUs)

Under the Gramin Krishi Mausam Sewa, the IMD proposes to establish District Agro-Met Units (DAMU) in 530 districts, in addition to already operating 130 AMFUs, in order to meet the said expansion. Among other responsibilities, DAMU will receive weather forecast from IMD to prepare and disseminate sub-district level agromet advisory bulletins. Weather and climate information plays a vital role in agricultural production and management. In order to provide the weather information and agro-advisories needed for the block/taluka level farming community, India Meteorological Department (IMD) in collaboration with ICAR initiated to set up District Agro-Met Unit (DAMU) at each district across India under GKMS scheme. The Objectives of DAMU are;

- To prepare and issue in time, crop specific, weather-based District/Block level Agro-met Advisory for next 5 days by AMFUs and DAMUs.
- To collect and organize climate/weather, soil and crop information and to amalgamate them to prepare database for value addition of advisory
- To improvise the existing district level Agro met Advisory Services (AAS) so as to deliver crop and location specific AAS to farmers at block level.
- To design optimum observatory network for issuance of village level advisories for implementation of crop weather insurance.
- To establish District Agro met Units as nodal center for catering to needs of agriculture services.
- To provide customized advisory bulletins through last mile connectivity to farmers with personalized agro met advisory services.
- To extend the weather based advisory service to the allied areas like livestock, grazing of farm feed *etc.*
- To establish appropriate dissemination and support system for weather-based crop insurance in the country.

ICAR through All India Co-ordinate Research Project on Agro-meteorology (is pursuing R and D on Agro-meteorology through a network of 35 centers located with SAUs in the country for improvement in weather based advisory and strengthening outreach of advisory bulletin to the farming community. The central Research Institute for Dry Land Agriculture (CRIDA) and National Innovation on Climate Resilient Agriculture (NICRA) projects of the Indian Council of Agricultural Research aim to enhance resilience of Indian agriculture to clime change and climate variability through strategic research and technology. It also provides agro advisories to the crop weather outlook, website established by CRIDA and guides the Ministry of Agriculture on contingency planning during the crop season.

11.8 Use of Satellite in Weather Forecasting

A weather satellite is a type of satellite that used to monitor the weather and climate of the Earth. The Vanguard 2 is the first weather satellite was launched on February 17, 1959. The first weather satellite to be considered a success was TIROS-1, launched by NASA on April 1, 1960. There are two types of weather satellites;

a) **Polar-orbiting** satellites orbit at low-altitude around the North or South Pole and monitor the entire planet over the course of 10 days to 1 month. The information and data from these satellites are used in weather forecasting.

b) **Geostationary** satellites orbit at high-altitude at the equator, at the same speed as the Earth's rotation, continuously monitoring one section of Earth. These satellites can track cloud formations and large storms as well as fires, *etc.*

Indian National Satellite (INSAT) is domestic satellite system used in communication, television, radio broadcasting and meteorology. ISRO launched its first atmospheric observation satellite Bhaskara with SAMTR for the observation of an atmosphere and ocean in 1972. The Indian National Satellite (INSAT) system was commissioned with the launch of INSAT-1B in August 1983 (INSAT-1A, the first satellite was launched in April 1982 but could not fulfill the mission). INSAT-2A and 2B launched in 1992 and 1993, INSAT-2E launched in 1999. A geostationary meteorological satellite (METSAT) system devoted totally to meteorology was launched in 2002, it has been renamed as Kalpana-1. INSAT-3A has been launched in April 2003 and carries identical payloads as in INSAT-2E. Currently IMD uses geostationary meteorological satellite INSAT-3D (launched in 2013) and INSAT-3DR (launched in 2016) and polar satellite OCEANSAT-2, SCATSAT-1 and Mega-Trophiques.

The data from INSAT satellites are being used to retrieve a number of quantitative products. INSAT imagery is being used very exhaustively to provide support for synoptic analysis and weather forecasting. INSAT series give information on cloud spread and temperature of the earth surface and cloud tops.

This information is used in short range forecasting and medium range forecasting. They are also used for detecting cloud amount and intense precipitation, the eye of cyclone and cloud movement could be clearly seen through satellite picture which will help in predicting cyclonic movement.

11.9 Smart Phone Applications: Role in Weather Advisory Dissemination

Agriculture continues to be the most important sector of the Indian economy. Research, extension and farmers efforts have all contributed significantly in increasing food production. Agriculture and farming have been always a basic need for past, present and future. All human survival and culture flourish only when farming community is well developed. Prediction of weather in advance is known as weather forecasting. The India Meteorological Department, State Agricultural Universities, IITM, KVK, Government departments, AMFU, FASAL and DAMU together work on collection and preparation of weather advisory based on present and past meteorological data. The importance part of the weather forecasting is dissemination of the information to the each and every farmer of the country. The wide spread network of mobile phones could be the game changer in this problem. It will put agriculture field to its zenith. The mobile app is one of the platforms, where a farmer can get all solution and information in just one touch. During the past decades, agriculture information and technology transfers are mostly done by village level workers, extension personals, scientists, subject matter specialists of KVKs, universities *etc.* With the arrival of the internet, most of the information were tried to avail by e-based services. But it has reached only limited users due to installation cost of the computer devices, SMSs and voice message delivery systems are easy, but it requires a special type of options or formats to be sent to the system to get the precise information, therefore ICTs are moving in the direction of mobile apps.

Smart phones have been identified as one of those effective innovations which benefited a large number of people in the developing world. In India, mobile applications are transforming agriculture. Agriculture is prime sector of importance. More than 50 per cent of people are engaged in agriculture activity. To make agribusiness productive, smooth and respectable it is important that, it should be linked to recent technologies. These technologies need to be smarter, faster and cheaper to use. Mobile application one of such technology that can be used directly in agricultural growth. Although this channel of information dissemination is in juvenile phase but it's advantages can be seen in near future. The strategies for expansion of application-based information require expulsion of obstacles like better modest handsets, compatible smart phones, multilingual platform, subsidizes internet packs, regular trainings and awareness amongst farmers. Smart phones are the example of overcoming adversity of connecting the rural digital divide, bringing monetary advantages and acting as catalyst for social mobilization through improved communication.

Mobile based applications are nearly verge of replacing the computer-based services due to its cheaper cost and easy integration with various cellular services. Mobile based revolution is a package, which is led by smart phones, internet service providers and application developers. There are thousands of applications present today in the area of agriculture. Timely access to information is a crucial requirement for decision making in agriculture and allied sectors. Information and Communication technologies are facilitating faster access and exchange of information. Among ICTs there has been increasing use of mobile phones with a number of services provided by various agencies. With the increase in the number of apps for farmers, it is important to improve awareness on the specific information and services provided by these apps. Various mobile apps, which enable access to get location specific weather based agro advisory are given below.

11.9.1 MAUSAM

The MAUSAM mobile application is launched on the occasion of the 14th foundation day of Ministry of Earth Sciences on July 27, 2020 by Union Minister for Health and Family Welfare, Government of India. The aim is to improve the dissemination of weather forecast and warning services based on latest tools and technologies. This app is designed and developed by the International Crops Research Institute for the Semi-Arid Tropics, the Indian Institute of Tropical Meteorology, Pune and the India Meteorological Department. The Mobile app has been dedicated to the general public and is designed to communicate the forecasts and weather information in a lucid manner. Apart from the weather forecasts, users will also be able to access the radar images and will be proactively warned of impending weather events. A mobile app is an important tool for the dissemination of weather information and warnings in a user-friendly and attractive manner. The mobile-based application will provide weather forecasts, now cast and warning for various cities and locations.

It provides information on **Nowcast**: It will showcase a three-hourly warning of localized weather phenomenon and their intensity issued for districts of India and for about 800 stations. In case of serious weather, its impact also will be included in the warning. **Current Weather**: Under this, data on the humidity, current temperature, direction, and wind speed for 200 cities will be updated 8 times a day. Information on sunrise, sunset, moonrise, and moonset will also be provided. The app will provide the information on the past 24 hours and the 7 days forecasts of weather conditions in around 450 cities. **Warnings**: The app will also issue alerts twice a day for all the districts for the next 5 days in colour code (Orange, Red, and Yellow), in order to warn the citizens of dangerous weather. Red colour code will be the most dangerous category which will urge the authorities to take action, orange colour code will prompt the authorities and public to be on alert and yellow code will prompt the authorities and the public to keep themselves updated. The latest station wise radar products will be updated every 10 minutes. The Mausam mobile app will be an important tool for dissemination of weather information and warnings in an attractive and user-friendly manner which will meet the requirements of public. This mobile app is available both on Play store and Appstore.

11.9.2 MEGHDOOT

It has been developed by experts from the India Meteorological Department and Indian Institute of Tropical meteorology and the Indian Council of Agricultural Research and launched in 2021. It is mobile application that will provide location, crop and livestock-specific weather-based agro advisories to farmers in local languages. The app would provide information in the form of images, maps and pictures.

It will provide forecast relating to temperature, rainfall, humidity, wind speed and direction. The app plays a crucial role in agricultural operations and advisories to the farmers on how to take care of their crops and livestock. It has been integrated with WhatsApp and Facebook to help farmers by sharing advisories. The app provides district-wise advisories on crop and livestock management issued by Agro Met Field Units every Tuesday and Friday based on the past and forecasted weather information. It will help the farmers to take weather-sensitive decisions like sowing of crops, pesticide and fertilizer application, irrigation scheduling and vaccination of animals.

11.9.2 DAMINI

Damini App was developed by the Indian Institute of Tropical Meteorology, Pune and Earth System Science Organization under the ministry of earth sciences and Launched on the occasion of 57th Foundation Day Celebration at IITM in 2020. The app monitors the lightning occurrence all over India and alerts the user of lightning near them by a GPS notification under 20 km and 40 km and movement and direction of thunderstorm. Further, the Damini app also triggers warning about lightning strikes three hours in advance which can help reduce losses to life and property.

11.9.4 UMANG (Unified Mobile Application for New-age Governance)

It is launched in 2017 to bring major government services on a single mobile app, with a larger goal to make the government accessible on the mobile phone of our citizens. The Ministry of Electronics and Information Technology has added the Indian Meteorological Department weather services to the Umang Mobile App on 22nd May, 2020.

It will forecast 7 weather related services including; Current temperature, humidity, wind speed, direction for 150 cities updated 8 times a day. Information on sunrise/sunset and moonrise/moonset is also given. Three hourly warnings of localized weather phenomena and their intensity issued for about 800 stations, and districts of India by State Meteorological Centers of IMD. In case of severe weather, its impact also is included in the warning. Past 24 hrs and 7 days forecast of weather conditions around 450 cities in India are given. District rainfall information daily, weekly, monthly, cumulative series are available. Past 24 hours and 7 days forecast of weather conditions of around 100 Tourist cities in India are provided. The alert issued to warn citizens of approaching dangerous weather. It is colour coded in Red, Orange and Yellow are the alert levels with Red as the most severe category. Issued twice a day for all districts for the coming five days. Cyclone warnings and alerts provides the track of cyclonic storms along with likely time and point of crossing of coast. Impact based warnings, area/district wise, are issued so that appropriate preparation including evacuation of vulnerable areas can be done.

11.9.5 BHUVAN HAILSTORM

It is developed by the Ministry of Agriculture and Farmers Welfare, Govt. of India. A mobile app has been developed to capture crop loss happened due to hailstorm. Agriculture Officer will go to the field with mobile or tablet loaded with this mobile app. This mobile app is able to capture following parameters: Photograph of field with latitude and longitude, name of crop, date of sowing, date of likely harvesting and source of irrigation. This captured data will automatically be plotted to Bhuvan Portal and analysis can be done easily.

11.9.6 Kisan Suvidha

It is developed by the Ministry of Agriculture and Farmers Welfare, Govt. of India. Kisan Suvidha is an omnibus mobile app developed to help farmers by providing relevant information. The app provides information to farmers on weather, market prices, dealers, plant protection, IPM practices, seeds, expert advisory, Soil Health Card, godowns and cold storage. The information is currently provided in English, Hindi, Tamil, Gujarati, Odia and Marathi.

It includes information on; Weather: It includes details of humidity, temperature, wind and rainfall for the current day and the forecast for the next five days. Unique features include extreme weather alerts like hailstorm or unseasonal rains. Market Price: Provides latest market price of commodities in the nearest mandi and the maximum price in the district, state as well as in India. Prices are given of crops traded in a registered agriculture market or mandi of the particular district. Agro Advisory: Agro advisory section shows messages for farmers from agriculture officials and state universities in the local language. Plant protection: This section gives pest, weed and disease-related information as well as management practices for each stage of crop development from Seedling/nursery to harvesting. One can also upload a picture of the affected crop and get a response. Soil Health Card: Information pertaining to soil health is available for farmers who have registered. Dealers Information: Provides information on seed, pesticides, fertilizer, farm machinery dealers. Call to KCC: The app also directly connects the farmer with the Kisan Call Centre where technical graduates answer farmers 'queries. Cold stores and Godowns: Information on cold stores and godowns is also added.

Chapter 12

Agriculture and Weather Relations

12.1 Agriculture and Weather Parameters

Solar radiation, temperature, air pressure, wind, atmospheric humidity, cloudiness, precipitation are the elements of weather. These are important factors on which the growth, development and yield of a crop depends. The study and effect of these elements can be used in agriculture for planning, stability and management of the crops, arranged agriculture operation on weather basis. Knowledge of agro-meteorology and the impact of weather and climate on crops will greatly help to devise appropriate crop production strategies for taking advantage of favorable climate and weather conditions and avoiding risks due to their hazards for realizing optimum and sustainable agricultural productivity. Weather during crop season strongly influences the crop growth and development and accounts for two thirds (67 per cent) of variation in productivity, while the crop management practices account for one third (33 per cent). Climate factor and an understanding of the complex processes of interaction between climate and biological processes of plants are essential to a scientific approach to farming, based on planned cropping pattern and improved land and water management practices.

12.1.1 Light

Light is the primary source of energy, without which living organism could not exist. It is a vital factor to all living things. Green plant obtains the energy necessary for life directly from sunlight. The amount of light received by plant is determined by intensity of light and its duration. In natural environment the length of day may have an even greater effect on the total amount of light received than does light intensity. The light intensity affects the crop plants through its effect on photosynthesis. Very low light intensity reduces the photosynthesis as stomata are closed which restrict the intake of CO_2 gas.

The rate of photosynthesis increases with increase in light intensity. The light intensity at which the photosynthesis intake of CO_2 is equal to the respiratory output of CO_2 is called compensation point. Therefore, net photosynthesis is zero at the compensation point. The light intensity at which respiration equals photosynthesis is known as light compensation point. The light intensity at which rate of photosynthesis becomes highest is known as light saturation point. The change in the morphology of plants due to light is called as photomorphogenesis.

The photosynthesis process of plants occurs in visible portion of solar spectrum. The visible portion of solar spectrum having wavelength range from 0.4 to 0.7 μm is known as photo synthetically active radiation (PAR). The red spectrum (580 nm to 680 nm) of light which is the most effective for inducing flowering in both short day and long day plants. Green plant exposed to monochromatic light *i.e.* >680 nm in red zone, this prolonged duration of red zone decreased the rate of photosynthesis, this effect is called red drop. The quantum yield produced by the mixed light was greater than the total yield got from the two beams of light used separately, this enhancement effect is known as Emerson effect.

Photoperiodism is the response of plants to relative length of day and night;

- ✰ Long day plant: Plant flowers normally when photoperiod is greater than 12 hours *e.g.* Wheat, Barley, potato and sugarcane.

  The plants flowers strictly only when the plants are exposed to the required photoperiod is known as long day obligate plants. *e.g.* carnation, oat.

  The plants flowers in any photoperiod are known as long day facultative plants. *e.g.* wheat, pea.

- ✰ Short day plant: The plant flowers normally when photoperiod is less than 12 hours *e.g.* Tobacco, soybean, maize.
- ✰ Day natural plant: Not affected by the photoperiod *e.g.* Tomato, cotton, pineapple.

Phototropism is the orientation of a plant in response to light.

- ✰ Positive phototropism: It is the orientation of a plant in response to light towards the sources of light *e.g.* sunflower.
- ✰ Negative phototropism (Skototropism/Photonasty): It is the orientation of a plant in response to light away from the sources of light.

Photosensitive plants whose reproduction is influenced by variation of length of photo period *i.e.* the period of exposure of sunlight and can be grown in certain season. *e.g.* Mayflower.

Photo insensitive plants whose reproduction is not influenced by photoperiod and can be grown round the year. *e.g.* All rice varieties.

12.1.2 Temperature

Temperature of day and night also affects the crop. If the days are warm, good for photosynthesis but if nights are warm, the loss of respiration is higher. High night temperature favors growth of shoots and leaves and also affects plant metabolism. Higher temperature disturbs the physiological activities like photosynthesis and respiration. Higher temperature increases respiration leading to rapid depletion of reserved food. It also increases transpiration loss of water and causes water deficit in crop plants under limited water supply.

The response of living organisms to regular changes in temperatures is known as thermoperiodism. The temperature range of **10-40°C** is the favorable range for crop plants. The plant cells get killed when they are exposed to the temperature range from 50 to 60°C. This point of temperature called thermal death point. This varies from plant to plant and species to species. Cardinal temperature is the maximum and minimum temperature that define limits of growth and development of plant. Optimum temperature is the temperature at which growth and development of plant/crop is best carried on. The minimum, maximum and optimum temperature of individual's plant is called as cardinal temperature.

Crops	*Minimum Temperature °C*	*Optimum Temperature °C*	*Maximum Temperature °C*
Rice	10-12	30-32	36-38
Wheat	3-4.5	20-25	30-32
Maize	8-10	20-32	40-43
Sorghum	12-13	32-35	40
Tobacco	12-14	29	35
Sugarcane	18-20	26	42
Cotton	15	32	43
Groundnut	15	25	40
Sugar beet	4-5	25	28-30

12.1.3 Moisture

An abundance of moisture results in a rich natural flora and makes possible a wide choice of crops. Deficiency of moisture, permits only narrow range of potential crops, and is accompanied by hazards to efficient crop production. Functioning of stomata is often greatly reduced under low moisture/moisture stress condition. Photosynthesis is hampered by reduction of stomatal opening limiting CO_2 absorption. Cell size and shape are reduced under moisture deficient condition and ultimately plant growth and yield are reduced. Too much water may be harmful to plants just as too little. High soil moisture and humidity condition cause failure in establishing stand of legumes and grasses. Excess of soil moisture in earlier growing season may result in more vegetative growth and delayed flowering.

12.1.4 Relative Humidity

Water is present in the atmosphere in the form of invisible water vapour, normally known as humidity. Relative humidity is ratio between the amount of moisture present in the air to the saturation capacity of the air at a particular temperature. If relative humidity is 100 per cent it means that the entire space is filled with water and there is no soil evaporation and plant transpiration. Relative humidity influences the water requirement of crops. Increasing relative humidity decreases the rate of transpiration. Reduction in transpiration reduces the translocation of food material and uptake of nutrients. The relative humidity ranges from 60 to 80 per cent is good for crop growth. A very high relative humidity is beneficial to maize, sorghum, sugarcane and it is harmful to gram, sunflower and tobacco. Humidity also affects the water requirement through evapotranspiration. High humidity with high temperature favours the outbreak of pest and disease.

12.1.4 Precipitation

Precipitation includes all water which falls from atmosphere such as rainfall, snow, hail, fog and dew. Rainfall one of the most important factors influences the vegetation of a place. Total precipitation in amount and distribution greatly affects the choice of a cultivated species in a place. In heavy and evenly distributed rainfall areas, crops like rice, tea, coffee and rubber are grown. Low and uneven distribution of rainfall is common in dry land farming where drought resistance crops like pearl millet, sorghum and minor millets are grown. In desert areas grasses and shrubs are common where hot desert climate exists. Though the rainfall has major influence on yield of crops, yields are not always directly proportional to the amount of Precipitation as excess above optimum reduces the yields. Distribution of rainfall is more important than total rainfall to have longer growing period especially in dry lands.

12.1.5 Wind

The horizontal movement of air is called wind. Wind increases the transpiration, intake of carbon dioxide and photosynthesis of plants. Wind influences the plant life both physiologically and mechanically. Wind affects plant directly by increasing the transpiration and intake of CO_2. Thus, raising the supply of CO_2 to the plants and thereby increase in photosynthesis. However, the increase in photosynthesis is again up to a certain wind speed beyond which its rate become constant. The warm and dry winds increase evaporative loss and cause moisture stress when soil temperature is very low. Lodging is one of severe injury caused by strong winds which causes great yield loss depending upon the stage of crop. This injury is common in paddy, maize, wheat.

12.1.6 Phenology

It is the study of periodic biological phenomenon. It qualitatively describes the successive stages in the development of plants, from seed germination to flowering to maturity. The crop growth stages also known as phenophases. The environmental and cultural factors directly or indirectly influencing crop phenology.

12.1.7 Vernalization

The term vernalization was coined by T. D. Lysenko (1928) for the effect of low temperature on flowering in plants. vernalization is the promotion of flowering in response to prolonged low temperatures. It is the induction of a plant's flowering process by exposure to the prolonged cold of winter, or by an artificial equivalent. After vernalization, plants have acquired the ability to flower, but they may require additional seasonal cues or weeks of growth before they will actually flower. The low temperature requirements for flowering called vernalization. The optimum temperature for vernalization is assumed to be in the range of 0-7°C, with temperature between 7 to 10°C having decreasing influence on the process.

12.2 Weather Parameters and Insect-Pests

12.1 Temperature

It is the degree of hotness or coldness.

Solar radiation is the main source for air temperature. The insect ecology is the scientific study of how insects, individually or as a community, interact with the surrounding environment or ecosystem. Ecosystem is the basic functional unit of ecology, is composed of abiotic components of the environment and biotic community both interacting with each other. Abiotic components are the non-living aspects of the environment that affect an organism, such as temperature, moisture and humidity, rainfall, light, atmospheric pressure, air currents, water, oxygen, pH, salinity, place to live *etc.*

Temperature is one of the important variables, which influences all stages of crop and insect-pests activity during its growth development and reproductive phase. Based on the maintenance of body temperatures, animals are generally divided into; Warm blooded also known as homeothermic and endothermic animals which maintains the body temperatures constant irrespective of atmospheric temperatures. *e.g.* Mammals, Birds. Cold blooded also known as poikilothermic and ectothermic animals which change their body temperature when the atmospheric temperature changes. *e.g.* Insects. Socio-homoeothermic animals such as honey bees can able to maintain their body temperatures slightly above the atmospheric temperatures, and are able to air condition their nests. They maintain their own temperatures inside their colony irrespective of the temperature outside. Insects have been considered as poikilotherm animals which means animals in which body temperature is variable and dependent on ambient temperature as opposed to being homeothermic animals. However, the term thermoregulation, is currently used to describe the ability of insects and other animals to maintain a stable temperature either above or below ambient temperature, at least in a portion of their bodies by physiological or behavioral means. While many insects are ectotherms, the animals in which their heat source is primarily from the environment, others are endotherms, the animals which can produce heat internally by biochemical processes. These endothermic insects are better described as regional heterotherms because they are not uniformly endothermic. When heat is being produced, different temperatures are maintained in different parts of their bodies, for example, moths generate heat in their thorax prior to flight but the abdomen remains relatively cool. Temperature directly influences the distribution of insects in nature, because certain amount of warmth is essential for growth and reproduction. This is the most important physical factor of the environment. It affects directly on the dispersal, distribution, movement, development, metabolism, fecundity and reproduction which determine the abundance of the insect and also indirectly through influence on food availability and other environmental factors such as moisture, air movement *etc.* The various factors of Insects influenced by temperature are as follows;

Insect Development

The highest temperature limit at which the insect just lives and dies beyond it, is known as a vital temperature maximum while the lowest temperature at which the insect can live but may not exhibit any activity is called the vital temperature minimum. *e.g.* stored grain pest *Rhizopertha dominica* dies at a temperature of 62.8°C exposed for 5 minutes. Hibernating insects can survive at a very low temperature of -50°C. In general, it is convenient to discuss the effect of temperatures in three zones *viz.*, zone of effective temperature, zone of inactivity and lethal or fatal zone. The growth of cold-blooded animals is arrested at 10°C and this temperature is called developmental zero or threshold development. The optimum temperature for the normal development of insects is 10-35°C is known as zone of optimum or normal development. Active development takes place in zone of effective temperature zone. The zones immediately below and above the threshold of development, and in these zones the insect will live but without development, and they can recover if removed to favorable temperatures known as zone of inactivity (-5°C to 10°C, and 35°C to 50°C). Zone of fatal temperatures (-5°C to -14°C, and 50°C to 60°C) is the zone fatal for insects, and cannot survive and die.

The optimum range of temperature depends on the specific species. It also depends on the physiological state of the individual insect. Some insects can tolerate the lower temperature during fall and winter than in the spring or summer. Insects of tropics tolerate less of cold than those of temperate areas. If the insect is given a choice to move along a temperature gradient, it prefers a narrow limit of temperature known as temperature Preferendum or preferred temperature. Certain insects do not get freezes even the surrounding temperatures go below the freezing point known as cold hardiness/super cooling. These insects have cryoprotective compounds like glycerol, sorbitol and erythritol which help insects not forming ice crystals in haemolymph by depressing the freezing point. The freezing point of most insects is between -10°C to -2°C. Under unfavorable temperatures, insects suspend their activities known as diapause, and these are two types; Hybernation is the period of suspended activity occurring during seasonal low temperatures *i.e* winter, while Aestivation is a period of suspended activity occurring during seasonal high temperatures or in dry season *i.e.* summer. This reversible inactivity can last for number of weeks during winter and summer, respectively. As soon as the temperature becomes moderate, these insects resume activity.

Fecundity/Reproduction

The insect's fertility its development ceases beyond the tolerable temperature range for reproduction and development. Within the tolerance range the development of eggs, the rate of oviposition, and larval and pupal development normally increase with the increase in temperature. *E.g.* Pediculus which does not lay eggs below 25°C and *Tenebrio molitar* which shows a considerable decrease in the pupal period with the increase in temperature within tolerable range.

Fecundity is the egg laying ability of the insects. In insects, fecundity is maximum at moderately high temperatures and decline at both upper and lower limits of favorable temperature. Aphids remain parthenogenetic under high temperatures and many hours of sunshine while the opposite condition gives rise to oviparous forms. Within the optimum temperature range during which the reproduction and development ranges are maximum. Under fluctuating temperature, the development rate is higher than under constant temperature, *e.g.* eggs of grasshoppers. Such a situation is common under field conditions.

The moulting process in insects is also affected by temperature. In some insects the number of moults increases at high temperature whereas in other it decreases. *E.g.* at a temperature of about 22°C, 6 moults are recorded but at 32-37°C there are only 7 moults by grasshopper. Grasshopper lays 20-30 times more eggs at 32°C compared to 22°C. Oviposition of bed bug inhibited at 8-10°C. Early shoot borer of sugarcane attacks higher temp. Larval period of sugarcane internode borer very short 16-24 days in summer, prolonged 141-171 days in winter.

Insect Distribution

The distribution of insects is affected by temperature particularly in low temperature zones. The distribution limit of insects in the northern extreme in temperate zones is determined by the low temperature extremes. Below the extreme distribution limits there is a zone which is densely populated during summer but during winter the insects usually die. The insects are abundant in high temperature zones because of variations in the temperature. As insect in north has less number of generations in a year as compared to the insect of southern zone. Mediterranean fruit fly *Ceratitis capitata* could not establish in England and North Europe, because it cannot withstand temperature below 10°C. Mosquitoes are abundant at 10°C, but are rare at 30°C.

Dispersal

The rate of dispersal of insects is dependent on the temperature. Insects are more active in tropical zone than in temperate zone. Insects tend to move away from unfavorable temperatures. The rice weevil *Sitophilus oryzae* found in the upper layers of storage bins, because the inner layers are warm and it cannot tolerate beyond 32°C and hence adults migrate to upper layers of the bins. Desert Locust *Schistocerga gregaria* start gathering in basking groups to gain warmth and they take flight only when the temperature near the ground is between 17-22°C and stop migrate when the temperature goes down to 14-16°C.

Acclimation

Acclimation is the process in which an individual organism adjusts to a gradual change in its environment such as a change in temperature, humidity, photoperiod, or pH, allowing it to maintain performance across a range of environmental conditions.

a) *Low Temperature Acclimation*

The insects have been found adapted in the arctic by showing cold hardiness. Certain insects like *Aedes aegypti* can survive at low temperatures for a longer period and even show acclimation to cold but die due to freezing of the body fluids. Some insects can withstand super cooling thus can prevent freezing. The presence of some cryoprotective compounds in haemolymph such as sorbitol, glycerol and erythritol which depresses the haemolymph freezing point are helpful in freezing tolerant insects. During their inactive stage, the glycerol appears in the tissue and disappears soon after this state is over.

b) *High Temperature Acclimation*

It is useful to insects living under natural circumstances of daily high temperatures. It has also been observed in certain insects like Calliphora and Phormia. At high temperatures, locusts expose minimum body surface to sun rays by lying parallel to them while they expose maximum body surface to sun rays at low temperatures laying at right angles to them.

Temperature and Consumption of Food Reserves

When only limited quantity of food is available, the rate of consumption of food reserves is also dependent on the temperature. This in turn affects the survival of the insect. For example, *Glossina* spp. fly depends on food reserves between two meals. If after one meal the food reserve is completely consumed and the fly does not get another meal immediately it will die.

Insect Body Temperature

Sun radiations and metabolic rate are major sources of heat gain. It is due to solar radiation that the temperature of the insect body becomes different from the ambient temperature. The different factors like size, color, and shape of the insects and orientation with respect to sun, influence the effect of solar radiation on the insect body temperature. A larger surface area having insect is more affected than the smaller insect. The insects of lighter colours absorb less radiations, than the dark ones. Heat loss in an insect is caused by various processes like evaporation, convection, conduction and long-wave radiation. In case of solar radiation, the metabolic heat is the only source of heat during movement of insects. In the absence of solar radiation, the major loss of heat from the body takes place by evaporation. The rate of evaporation is dependent partly on the size of the insect. A smaller insect loses more water than the larger insect as the smaller insect has a larger ratio of surface area to volume.

The movement of air also increases the heat loss as a steep gradient is maintained on insect and its surroundings. Hairs and scales present on the body of insects prevent the heat loss. In the event of a fall in surrounding temperature the insect maintains its body temperature by active body movement and muscles contractions. *E.g. Sphingididae* family moths the body is kept warm by continuous wings movements. The activity of insects is directly proportional to the body temperature. When the body temperature falls, the metabolism slows down and its activity retards whereas on rise in the body temperature, the insects become more active until at a certain temperature it becomes inactive. A further rise in temperature causes death.

12.3 Relative Humidity

Humidity is the amount of water vapour present in the air which depends on the atmospheric pressure and temperature. The water in vapour form plays important role in life of terrestrial organisms. A certain amount of atmospheric humidity is essential for controlling the loss of water from body. If the humidity is low, the animals die from desiccation at high temperature, and if the humidity is too high, it may harm in many ways, including the development of fungi and bacteria epizootics. Water is essential constituent of protoplasm, *i.e.* life, though the actual quantities vary in different insects. *E.g. Tribolium, Sitophilus, Trogoderma* and *Callosobruchus* are the stored grains pests; water constitutes 50 per cent of their body weight. However, bodies of most of insects contain 80-90 per cent water and larvae of certain dipteran that live in moist surroundings contain up to 98 per cent water.

In nature, insects are found under a wide range of moisture conditions from fresh water to the driest sun-dunes in deserts. In aquatic animals, the dryness is expressed in terms of osmotic pressure of the water. The truly terrestrial insects can live in dry places under varying conditions of moisture. *Collembola* live in very humid places where air is always saturated with moisture. At the other extreme, the desert insects, which can survive in air containing less than 10 per cent moisture. Food is the source for all insects, their feeding habits reflects adaptations to cope with conditions of excessive moisture or shortage of water *e.g.* aphids and bugs, ingest large quantities of plant sap and get rid of excessive water through excretion and also special arrangement in digestive system (filter chamber). Those living on dry foods such as grains do not get sufficient water *e.g. Ephestia* larvae could live in flour dried at 103°C. This led to discovery that these insects could retain body water through metabolism. The fall of water content of body below certain minimum proves disastrous to insects and if it is considerably above the normal (in wet places) harmful effects like viral, fungal and bacterial disease outbreaks occur. If the insects are given a choice to move freely in a humidity gradient with complete dryness on one end and the saturated air on the other, they have tendency to congregate within narrow range of humidity and that range is the preferred humidity. The humidity influences the rate of development, fecundity and color *etc.*, in different insects. The various factors of Insects influenced by moisture are as follows;

Body Water Content

Water is an important for life. The different insect's metabolic activities are dependent upon water. A suitable balance is always maintained between the water content of the internal environment of the body and the moisture present in the external environment of atmosphere which includes humidity and rainfall. In insect body the water content varies from insect to insect from less than 50 percent to more than 90 percent of total body weight. Caterpillars comparatively have larger amount of water whereas the insects with hard bodies have lesser amount of water. More water is present in the active stages in insects in comparison to inactive stages.

Water Source

The main source of water is food for most of the insects. The quantity of water depends on type of food intake by insects. Those feeding on the succulent tissues of the plants have sufficient water whereas those that feed on dry food have little water. For arid zone insects the main source of water is obtained from metabolic process in the body. Some insects obtain water by absorbing it from the soil and atmosphere. The terrestrial insects living in dry areas obtain water by oxidation of the food within the body. Even fat is oxidized to produce water under certain conditions. The water from metabolic wastes is also utilized by *Tenebrio molitor*.

Insect Mortality

Insect mortality is dependent upon the moisture in surroundings. There is always an optimal range of moisture which is suitable for insects. Excess of optimal range kill the insects. At extreme low moisture the active stages of insects in particular are killed. Mortality also occurs at extreme high moisture. The death at extreme low temperature is due to excess loss of water and at high temperature the causes for death are variable, such as spread of viral, fungal and bacterial diseases *etc.*

Longevity

Environmental moisture determines the survival and length of life of insects. It is adversely affected by excess moisture. For example, *Locusta migratoria* live longer at lower humidity. Termites prefer high humidity 90-95 per cent relative humidity. Low relative humidity in rainfed groundnut crop induces leaf mines incidence.

Feeding Behaviour

At extreme moisture content in the environment the feeding activity of the insects is hindered. The larvae of spruce budworm stop feeding in water saturated air. The Tsetse fly does not feed at a relative humidity of 88 per cent.

Reproduction/Fecundity

Insect reproduction is also affected by moisture presence or absence. Excess moisture contents affect the reproduction. The Oviposition rate is adversely affected by low humidity. Infertility is shown by newly emerged adult migratory locusts, when relative humidity is below 40 percent. The number of eggs laid by a female increases as the humidity increases up to 70 per cent and declines again at 90 per cent. On the other hand, fecundity of the rice weevil, *Sitophilus oryzae* increases steadily up to 70 per cent humidity, but there is no decline in the number of eggs laid, if the humidity is increased. Coconut rhinoceros beetle *Oryctes rhinoceros* develops dark chitin in moist air and light color chitin in dry air conditions.

Insect Development

Insect development and moisture relation is specific to species. The high moisture content decreases the rate of development which may even be completely halted. On the other side the silkworm larvae do not pupate at higher humidity. The incubation time for eggs is reduced at higher humidity for example in spider beetle. The developmental stages prefer higher humidity. At an unfavorable humidity, the insect dies before completing its development. Even within the favorable range, all insects do not react the same way. There are three types: those in which the speed of development is retarded at high humidity (the nymphs of Locusta and Schistocerca), those have an accelerated development at high humidity (eggs of Locusta and Musca), those which have a rate of development independent to humidity (the embryonic development in the eggs of Bruchus not influenced by humidity). Generally, if the water content of the body is high, dry air accelerates the development. The adults of *Locusta migratoria* and *Schistocerca gregaria* become sexually mature quicker at a certain RH (about 70 per cent) than in atmosphere too dry or too humid.

Insect Adaptations to Conserve Water

Among terrestrial insects there are a number of other adaptations help them overcome unfavorable conditions of aridity or excessive moisture, and such adaptations may be morphological, biological and physiological.

A) Morphological Adaptations

- **Body pigmentation**: In colder regions, and on the mountains, insects usually have darker body pigmentation; whereas those in deserts are generally lighter in colour. Dark colour helps to absorb the sun light which raises the body temperature, helps in removing the excessive moisture in the insect body. The lighter body colour of desert insects helps in saving the insect from moisture loss through light reflection.
- **Integument:** A tough integument with fused sclerites and shield like pronotum of many coleopterans helps in water/moisture conservation, since most of them live in dry habitats and feed on dry foods. Waxy coating on body of some insects also helps in water conservation.
- **Pilocity (Hairyness):** A dense growth of hair on the body is common in many desert insects, which saves them from excessive evaporation and also aids in reflecting the sun light.
- **Winglessness:** Grasshoppers and Crickets which are commonly found in arid regions and semi-deserts have generally poorly developed wings and some are wingless. For example, the Phadka Grasshopper in Western India shows variation in size of wings and the Deccan Grasshopper is wingless altogether. In the desert areas of Africa, about half of the Orthoptera found are wingless. In some species of Tenebrionid and Carabid beetles, the wings are fused. This adaptation protects the insect against the effect of strong hot desert winds.
- **Form of the body:** The oval and compressed form of the body found in number of desert beetles allows the least amount of body surface being exposed to the surroundings. Moreover, they can seek shelter under stones, creeks and crevices.
- **Other characters:** A number of cicindelid and carabid beetles found in the desert have long and slender legs, with help of which they can walk on hot sand, without touching their bodies to sand. Other desert insects have strong legs for burrowing in sand, so they can burrow and bury themselves at faster rate, thus, escaping the desiccation by reducing the period of exposure.

B) Biological Adaptations

In general, the life-cycles of insects are in tune with environmental conditions. In summer, many delicate insects enter aestivation or are in diapauses. The pupa of Red Hairy Caterpillar, *Amsacta moorei* is in diapauses during autumn and winter and even part of next summer when dry conditions prevail. With the first shower in June/July, the pupae absorb moisture which is essential for breaking diapauses, and moths emerge and lay eggs. Groundnut red hairy caterpillar *Amsacta albistriga* (Arctidae-Lepidoptera) enter into aestivation when dry conditions prevail. Similarly, there is a diapause in summer in number of grasshoppers.

C) *Physiological Adaptations*

Cryptonephric condition for re-absorption of water from products of excretion so as to conserve water is present in meal worm Tenebrio Molitor and many other coleopterans and lepidopterans.

12.4 Light

The sun is the chief source of radiant energy and light on earth. The rays in the light spectrum play an important role in the biological field. For instance, the basic process of food manufacture in nature is accomplished under the visible rays, especially the blue and red. Wavelengths of 400-760nm are in visible range, and all the shorter wavelengths of radiant energy up to and including visible range, which is measurable is considered as an ecological factor of light. Different properties of light that influence the organisms are illumination, photoperiod, wavelength of light rays, their direction and degree of polarization. Intensity of light is measured in LUX, and luminosity is expressed quantitatively in terms on lumens.

The effect of light on the distribution and multiplication of animals is different from the effect of temperature and moisture, in that there is no lethal effect of low or higher levels of light. There are several species of insects which can live in total darkness or continuous light. Thus, for insects, there is no medium range of favorable light comparable with temperature and moisture. The importance of light lies in relation to the behavior of insects and the token stimulus which it provides to enable them to regulate and synchronize their life-cycles with seasonal change. Factors of Insects influenced by light. Light has a great influence on the different factors of Insects like the swarming, egg laying, changes in flight, prolongation of life, the emergence of many species, stimulating effect on insects.

Insect Movement

Light controls the locomotory activity of many lower organisms by direct action on their speed of locomotion. The higher activity under increased illumination is of great importance in the lives of many aquatic invertebrates and the smaller terrestrial forms, including insects. This undirected movement in relation to light is called photokinesis.

Insect Orientation

Phototaxis: Light also plays an important role in the orientation of animals through directed movements called phototaxis. The movement of an organism towards or away from source of light is called phototaxis. Phototaxis may be negative or positive. This behavior may serve to bring the animals to the right place at the right time. The phototactic reaction of animals to light can be modified or reversed by number of factors such as temperature, moisture, food and age. Green leaf hoppers of paddy *Nephotettix virescens* (Cicadellidae: homoptera) attracts to light in hot humid conditions. Colorado potato beetles (*Leptinotarsa spp.*) are ordinarily attracted to light but are repelled when the conditions are dry.

Insects which are active during day are called as diurnal, insects which are active during night called as nocturnal, and others which are active during dusk are known as crepuscular. For example, cockroaches are an example of a negatively phototactic organism, while moths are positively phototactic or phototactic, meaning they automatically move toward light. In honeybees, there is a correlation between hours of sunshine and their activity. Several Moths, leafhoppers, and beetles are known to be attracted to light at dusk and during night. Chafer beetles and many moths pass the day in concealment. Most of the moths are attracted to light and are known as photopositive or phototropic. This behavior has been extensively used through light-traps for observing brood-emergence (Rice stem borer) and the population fluctuations of some insect pests. Light-traps of different colours have been tried and found that most insects attract to UV light more than other lights. Moths are more sensitive to some wavelengths of light-ultraviolet, for example- than they are to others. A white light will attract more moths than a yellow light. Yellow is a wavelength moths don't respond to.

Cockroaches hide during daytime. Dusk (evening light) is most usual time for flight and copulation for moths. Dusk is most usual time for emergence of winged white ants. Grubs of Trogoderma spp. show -ve reaction to light (photonegative). Some insects are able to detect polarized light. The honeybee can perceive the degree of polarization of light in different regions of the sky and used this information in combination with other factors to determine the proper line of flight on a cloudy day. Based on this principle, artificial light can be employed as a source for attracting of deterring insects, in IPM programs. The traps with light source for attracting various insects known as light traps.

Light Hours

Photoperiodism: The number of hours of light in a day length of 24 hours is termed as photoperiod, and the response of organisms to the photoperiod is known as photoperiodism. Based on the response to light hours, the insects are classified in to two types; short-day species, which likes short day lengths, and when exposed to long day lengths, they enter into inactive phase *i.e.* diapause (Silkworm, Bombyx mori), long-day species, likes long day hours, and when exposed to shorter day lengths, they enter into inactive phase i.e diapause (cotton pink bollworm, *Pectinophora gossypiella*).

The motor activity rhythms of insects that are influenced by photoperiod include locomotion, feeding, adult emergence, mating and oviposition. The moulting and growth of some insects may be influenced by photoperiod. The diapause is stimulated by the photoperiod. The feeding mating *etc.* are regulated by the daily circadian rhythms with dawn and dusk periods in between which functions as a biological clock. The larvae of *Plutella xylostella* (diamond backed moth) reared in laboratory with 9hrs day light (all other parameters constant) completed the development in a longer period than those exposed to 15hrs day light. The insect photoperiodic responses are known to control polymorphism characterized by the body form, pigmentation or mode of reproduction. The winged and wingless forms of some aphids are produced due to the influence of photoperiod. In reduced photoperiod, sexual forms are produced in aphids. Bean weevil (*Bruchus obtectus*) lays more eggs in total darkness than in light. The fly of *Dacus tryoni* lays most of its eggs during the day, whereas the codling moth (*Laspeyresia pomonella*) does so during the night. The photoperiod plays a major role in the ecological adaptations and phonological synchronization of insects with their food sources.

Light and Growth, Moulting and Fecundity

Mulberry silkworm, *Bombyx mori* develops faster in light than in darkness. Grubs of khapra beetle, *Trgoderma granarium* develops rapidly in light than in dark. Moths of spotted bollworm of cotton (*Earias vitella*) and red hairy caterpillar (*Amsacta albistriga*) lay most eggs during periods of darkness. The bean weevil lays eggs during total darkness. Light stimulates the oviposition in mantids and inhibits in *Periplanata* spp. The light stimulates egg laying in Praying Mantis and other insects whereas it inhibits Oviposition in of *Periplaneta*. Fruit flies lay eggs in dark. Lepidopterans like cotton bollworm, red hairy caterpillar oviposit in dark.

Light and Body Pigmentation

In dark areas, less pigmentation develops in insects, while light color develops in more light areas.

Feeding

The plant feeding insects have senses of different wave-lengths of reflected light for host location. The direction of movement is affected by the position of sun and degree of polarization of light in different parts of the sky. Many insects are dependent upon plants for their food. The light has indirectly impact on the insects by affecting both terrestrial as well as aquatic plants. In aquatic ecosystem the main source of food are plants, the amount of light affects the photosynthesis in plants, and releasing more oxygen, the aquatic insect life depends on the oxygen concentration in water to a very great extent.

12.5 Wind

Air currents, especially the upper air being strong, carries many insects like aphids and scales to far-off places and is an important factor for dispersal of small insects. Insects are influenced by both air and water currents. These are determined by geographic conditions of particular insect habitat. The insect mortality is caused by the air movement. Severe winds may cause mortality. The movement of air also kills the insects due to desiccation. However, when the humidity is high the movement of air is beneficial. The distribution of several insects such as leafhoppers, aphids and many more insects are also influenced by winds. Sometimes wind is helpful in their dispersal while sometimes many insects are also killed by the strong winds by moving towards unfavorable environment. Water currents determine the existence of insects in different areas. The insects adapt themselves to face the water currents, for example Caddis flies get attached to submerged objects by their cases. The mosquito larvae and some other aquatic insects fail to survive in moving water. The black fly larvae get attached to stones or other material in water.

12.6 Rainfall

Rainfall is closely related to humidity, and RH is dependent on rainfall. The total amount of RF and its distribution in time influences the abundance of insects in a tract. Rainfall is >12.5cm rain during summer (May-June) results in increase in soil moisture, which is not favorable to the cutworms, and hence are forced to come out of soil and fall a ready prey to their parasites and predators. On the other hand, if the rainfall is less than 10-12.5cm during summer, they remain protected in soil and there can be a pest outbreak in the next season. So, the pest outbreak can be forecasted based on number of wet days during May-June, like if the number of wet days is <10, there will be an increase in cutworm in the following year.

Desert locust doesn't lay eggs unless the soil has sufficient moisture, and further even if it lays eggs, the eggs will not hatch until the sufficient moisture is present in soil. Rainfall also plays an important role in movement of swarms of desert locust. Saturated condition of the moisture is unfavorable for the development of cotton spotted boll worm *Earias vitella*. Red pumpkin beetle *Rhapidopalpa foveicollis* (chrysomelidae-coleoptera) withholds eggs until it comes across moist soil. Rain induces emergence of most of the insects from soil, while ants, termites, red hairy caterpillar, root grub beetles *etc.* emerge out from soil after receipt of rains. Rainfall is essential for adult emergence of cutworms. Heavy rain washes aphids, diamond back moth. Intermittent low rain increases BPH and thrips.

12.7 Atmospheric Humidity

Generally, of little importance. Locusts show great excitement and abnormal activity about half an hour before the occurrence of storm when the atmospheric pressure is low. Drosophila melanogaster flies stop moving when put under vacuum.

12.8 Edaphic Factors

The abiotic factor soil is also called edaphic factor and such scientific study is called pedology or edaphology. For plants, soil provides mechanical anchorage as well as water and minerals for nutrition, and for insects feeding on plant parts inside soil, stages of life cycle in soil, the soil factors play an important role for the developments and distribution of insect pest populations. The nature of edaphic factors determines the nature of plant community which further influences the animal life forms; soil texture, soil air, soil PH, soil temperature, soil water. Many basic and applied studies in insect ecology have considered responses of insect populations to their physical or chemical environment. For insects that live above ground, the mechanisms of behavioral responses to environmental factors often are directly observable. However, behavioral responses of soil-inhabiting insects are much more difficult to observe and quantify. Soil texture and structure can have a direct impact on arthropod behavior and adaptations. Field studies of soil insects often quantify the consequences of behavior, whereas the actual behaviors are only inferred. There have been studies of ecological and physiological adaptations of several soil arthropods that are not considered to be agricultural pests but have certain life history features that make them amenable to investigation. However, most insects that are agricultural pests as soil inhabiting immature forms (*e.g.*, wireworms or rootworms in corn, scarab grubs) often have been studied in detail only in their more accessible, or observable, adult form.

Soil insects and other arthropods often move vertically in direct response to soil temperature, moving downward in late autumn to avoid freezing temperatures on the surface and returning to the surface in the spring to resume feeding. Certain soil insects, such as the black vine and strawberry weevils and white grubs, are favored by coarse-textured soils. Soil moisture is a major factor affecting the incidence, development, and control of pest problems. The length of white grub their life cycle is affected by prevailing soil temperatures in different climatic zones. The larval life is prolonged if the soil moisture is high. As the insect grows, less of the soil pore space is available to the insect for free movement, but it may take advantage of preexisting soil channels created by soil macrofauna such as earthworms, other arthropods, or small vertebrates. Water-filled soil pores can inhibit movement. Usually, eggs and pupae of insects are most resistant to soil moisture loss and least able to escape undesirable conditions. Larvae and adults are often mobile stages and may be able to move away or alter their behavior to minimize the impact of adverse moisture extremes. For example, scarab larvae can move downward in the soil profile to seek moister (and cooler) conditions during periods of heat or drought stress.

Heavily sclerotized soil arthropods may be less vulnerable to cuticular moisture loss than are less sclerotized forms, such as grubs or maggots. Legs of edaphic insects often are highly modified to facilitate digging or burrowing through soil. At least one leg segment is likely to be enlarged, specially shaped, and edged with spines or lobes to create functional spades. Expanded tibiae or femora provide increased surface area, providing improved leverage when the insect is moving soil. *E.g.* mole crickets use their greatly enhanced fossorial forelegs to burrow through soils very rapidly.

Chapter 13

Weather Hazards

13.1 Definition

A weather hazard is an extreme weather event that threatens life and property. It includes; drought, floods, frost, heat waves, cold waves, tropical cyclones, tornadoes, lightening, blizzards, earthquakes and tsunami. An extreme event is the occurrences of a value of weather variable above or below a threshold value near the upper or lower ends of the range of observed value of the variable. Oot It is a weather, climate or environmental condition that ranks above a threshold value near the upper or lower ends of historical measurements.

13.2 Drought

It is moisture deficit which results when the amount of water available in the soil is not sufficient to meet the demands of potential evapotranspiration. or it is a period of inadequate or no rainfall over extended period of time, creating soil moisture deficit and hydrological imbalance. Dry spell is a period of at least 15 conservative days during which not a single wetting rain has fallen. A drought year a year in which, the overall rainfall deficiency is more than 10 per cent of the long period average value. A drought prone area is the area in which the rainfall deficiency is more than 20 per cent than normal rainfall and rainfall deficiency is more than 40 per cent called as severe drought prone area. Aridity anomaly index is used to monitor the incidence, spared, intensification and recession of drought, monitor crop stress and used for crop planning and in the early warning system during drought/desertification. Aridity is the Thornthwaites concept to describe water deficiency experienced by plants. SPI (Standardized Precipitation Index) and NDVI (Normalized Difference Vegetation Index) are used to monitor status of rainfall situations.

13.2.1 Classification of Drought

a) Based on Source of Water Availability

i. **Meteorological drought:** It is the situation when actual rainfall is less than 75 per cent of the normal rainfall over an area.

a) **Slight drought:** It is the situation when actual rainfall is less than 11 to 25 per cent of the normal rainfall.

b) **Moderate drought:** It is the situation when actual rainfall is less than 26 - 50 per cent of the normal rainfall.

c) **Severe drought:** It is the situation when actual rainfall is 50 per cent less than the normal rainfall.

ii. **Hydrological drought:** It is the situation when hydrological sources like streams, rivers, reservoirs, lakes, wells dry up and ground water table depletes.

iii. **Agricultural drought:** It is the situation when soil moisture falls to meet the water demands of the crop during growth due to inadequate rainfall.

b) Based on Time of Occurrence

i. **Permanent drought:** Crop production due to insufficient rainfall in permanent dry or desert region is not possible without irrigation.

ii. **Seasonal drought:** It is due to large scale seasonal circulation that clearly defined rainy and dry climates.

iii. **Contingent drought:** It is due to irregular and variability in rainfall, especially in humid and sub humid regions.

c) Based on Medium

i. **Soil drought:** It is the condition when soil moisture depletes and falls short to meet potential evapotranspiration of the crops.

ii. **Atmospheric drought:** This results from low humidity, dry and hot winds and cause desiccation of the plants.

13.2.2 Strategy to Mitigate Drought

A) Farm Level

- ☆ Preventing and recycling of excess run off.
- ☆ Deep tillage to absorb and hold maximum moisture.
- ☆ Timely weed management to control water loss by ET.
- ☆ Planning for suitable cropping system.
- ☆ Selection of short duration varieties.
- ☆ Selection of drought tolerant crops.
- ☆ Adjusting sowing time to escape probability period of drought.

- ☆ Use of Mulching for conservation of soil moisture.
- ☆ Application of antitranspirant to reduce water loss through transpiration.
- ☆ Lifesaving irrigation.
- ☆ Reduction in plant population (thinning) to reduce ET

B) Govt. Level

- ☆ An early forecast of monsoon.
- ☆ Improved communication systems.
- ☆ Availability of resources like credit, fertilizers, pesticides.
- ☆ Adequate prices for produce in good year.
- ☆ Building reasonable buffer stocks for food grains.
- ☆ An improved transportation system.
- ☆ Crop Insurance.
- ☆ Proper assistance to farmers in the year when monsoon fails.

13.3 Floods

Floods are the major weather hazards, which are caused by heavy rainfall over a given area in a short period of time. These floods cause heavy damage to the crops and agricultural buildings also cause a greater loss of life and property than any other natural disaster. There are three types of floods; a) Flood is caused due to intense local rainfall: the rate of arrival of water at the surface of the earth is much greater than the infiltration rate into the saturated soil. In such cases, flash floods can occur. b) Flood occurs when snow begins to melt rapidly: This occurs only when the rise of temperature is associated with rain. In mountainous areas of mid and high latitudes, snow melt results flooding with the onset of the warm season. Such floods have severe agricultural effects. c) Winter flood: It is caused by rainfall lasting for many days. Although the rate of rainfall may be quite less but the total precipitation over a period of one or more days may be considerable. The fast-moving water droplets can damage the delicate crop plants and seriously affect their growth and hence yield is reduced.

Flooding can create anaerobic soil conditions that can have significant impact on vegetation. Chemical reactions in anaerobic soils also lead to a reduction in nitrate and the formation of nitrogen gas. This de-nitrification can be a significant cause of loss of plant vigour and growth following flooding. Along with this, heavy rainfall causes soil erosion, disruption to critical agricultural activities, water logging, increased moisture stress and attack of pests and diseases on the crops. Excess rainfall always causes water logging and runoff.

13.4 Heat Waves

World Meteorological Organization defines a heat wave as five or more consecutive days during which the daily maximum temperature exceeds the average maximum temperature by 5°C. There will be no harm to the human body if the environmental temperature remains at 37°C. Whenever the environmental temperature increases above 37°C, the human body starts gaining heat from the atmosphere. Heat wave is a period of abnormally high temperatures, more than the normal maximum temperature that occurs during the pre-monsoon (April to June) summer season. Heat waves typically occur between March to June and in some rare cases even extend till July. If the maximum temperature of any place continues to be more than 45°C consecutively for two days is called a heat wave condition.

In India, Heat wave is defined as the conditions over a prolonged period when the departure of maximum temperature from normal is +4°C to +5°C or more for the regions where the normal maximum temperature is more than 40°C and departure of maximum temperature from normal is +5°C to +6°C for regions where the normal maximum temperature is 40°C or less". Heat Wave is declared only when the maximum temperature of a station reaches at least 40°C for plains and at least 30°C for Hilly regions.

When the "departure of maximum temperature from normal is +6°C or more for the regions where the normal maximum temperature is more than 40°C and +7°C or more for regions where the normal maximum temperature is 40°C or less", it is defined as a severe heat wave condition.

High temperature results in the desiccation of the plants and disturbs the balance between photosynthesis and respiration, increase vapour pressure deficit and evapo transpiration. To minimize the impact, farmers are getting advice to apply frequent irrigations to the crops, mulches to maintain high moisture status in the soil and spraying of antitranspirants on crops.

13.5 Cold Waves

A cold wave is a rapid fall in temperature within a 24 hr period. A cold wave is determined by the rate at which the temperature falls and the minimum to which it falls. A cold spell is defined as the national average high temperature dropping below 20°F (–7°C). For declaring cold wave wind chill effective minimum temperature (WCTn) is used and when it is <10°C only, cold wave is considered. wind chill effective minimum temperature is the effective minimum temperature due to wind flow. IMD follows some criteria for declared clod wave;

- Normal: Minimum temperature is 10°C or more.
- Cold wave: Minimum temperature is 5°C to 10°C below normal.
- Severe cold wave: Minimum temperature is 7°C or more below normal.
- Normal: Minimum temperature is less than 10°C.
- Cold wave: Minimum temperature is 4°C to 5°C below normal.
- Severe cold wave: Minimum temperature is 6°C or more below normal.

Long and intensive cooling causes complete devastation of crops. To protect from cold injury, irrigation is important for standing crops as adequate soil moisture keeps the soil comparatively warm and saves it from cold and also smoke arrangement should be made around the fields and to cover the crop necessarily by green net *e.g.*, nursery vegetables, paddy, papaya *etc.* to save the crops from the ground frost. Grapes could suffer cracks due to the extreme cold conditions in the traditional belt of the crop Maharashtra. If mercury drops to 4.4°C, farmers are advised to arrange for smoking around the field and apply light irrigation to the crop and spray fresh water to the bunches.

13.6 Thunderstorm

It is a storms produced by a cumulonimbus clouds and always accomplished by lightning and thunder. Presence of warm and humid air in the lower layers of the atmosphere, atmospheric instability and intense convective activity are important requirements for the development of a thunderstorm. Thunderstorm is derived from the spanish word *tronar* means to thunder. The sound emitted by rapidly expanding gases along the channel of a lighting discharge is called thunder. A lighting stroke produces tremendous heat along its path. The air is heated which results in rapid expansion of the air column. The tremendous expansion of air sends vibrating pressure wave outside which moves at the speed of sound (3000 m/s), the noise is produce by such a wave. IMD follows some criteria for declared thunderstorm;

- ✰ Moderate: Loud peals of thunder with frequent lighting flashes, moderate heavy rains and maximum wind speed 29 to 74 kmph
- ✰ Severe: Continuous thunder and lighting, heavy rains and maximum wind speed 75 kmph

The cumulonimbus cloud consists of strong vertical updrafts and downdrafts. Depending upon their intensity (wind speed), they are classified as;

Weed Speed kmph	*Types*	*Weed Speed kmph*	*Types*
30-40	Gust wind	91-120	Moderate
41-60	Squall wind	121-149	Severe
61-90	Light wind	> 150	Tornado

13.7 Downburst

It is an area of strong winds produced by a downdraft. When it comes in contact with the earth, it spreads out laterally and causes damage at ground. There are two types of downbursts, a macroburst and a microburst. The macroburst covers an area larger than 4 kilometers in diameter, and the microburst is smaller than 4 kilometers in diameter.

13.8 Cloudburst

It is also known as a rain gush or rain gust is a sudden heavy downpour over a small region. The rate of rainfall equal to or greater than 100 mm per hour featuring high-intensity rainfall over a short period, strong winds and lightning.

13.9 Tornadoes

Tornado is a violently rotating column of air attended by a funnel shaped extending downward from a cumulonimbus cloud. Tornado is derived from the spanish word *tornada*. They are intense centers of low pressure having a whirlpool-like structure of winds rotating around a central cavity where a partial vacuum is produced by centrifugal forces. It is funnel shaped and its diameter is very small *i.e.* near about 10 meter and wind velocity is about 460 mile/hr. It is contact with both the ground and the cloud base. Where it touches the ground, there is complete destruction. Tornadoes are mostly experienced in United States of America. It originates due to mixing of dry polar air mass with warm and moist tropical air mass.

13.10 Hailstorms

Any thunderstorm which produces hail that reaches the ground is known as a hailstorm. Hail has a diameter of 5 mm or more. Hailstones can grow to 15 cm and weigh more than 0.5 kilograms. Hail forms in strong thunderstorm clouds, particularly those with intense updrafts, high liquid water content, great vertical extent, large water droplets, and where a good portion of the cloud layer is below freezing 0 °C. These types of strong updrafts can also indicate the presence of a tornado. The growth rate is maximized where air is near a temperature of -13°C. The storm's updraft, with upwardly directed wind speeds as high as 180 kmph, blow the forming hailstones up the cloud. The hailstone will keep rising in the thunderstorm until its mass can no longer be supported by the updraft. This may take at least 30 minutes based on the force of the updrafts in the hail-producing thunderstorm, whose top is usually greater than 10 km high. It then falls toward the ground while continuing to grow, based on the same processes, until it leaves the cloud. It will later begin to melt as it passes into air above freezing temperature. To minimize damage on crops farmers are advised to give mechanical support to the crops, harvest the matured crops and keep in a safe place.

13.11 Disaster

Disaster is an undesirable occurrence resulting from force that are largely outside human control, strike quickly with a little or no warning which causes serious disruption of life.

Classification of Disaster

- Atmospheric: Blizzards, lightening, tornadoes, cold wave, hot waves, thunderstorm, tropical cyclone, hailstorm, jet stream, frost, drought.
- Terrestrial: Earthquakes, landslides, volcanic eruption, soil erosion.
- Aquatic: Floods, tsunami, tidal waves, ocean currents, stream surge.
- Biological: Insect infestation, fungi, bacteria and viruses' diseases.

Chapter 14

Climate Change

14.1 Introduction

Climate change is any significant change in the measures of climate lasting for an extended period of time or it is any permanent change in weather phenomena from the normal of a long period average. Climate variability is a variations in the mean state and other statistics of the climate on all temporal and spatial scales, beyond individual weather events or it is the temporal changes in weather phenomenon which is part of general circulation of atmosphere and occurs on a yearly basis on a global scale. Climate variability looks at changes that occur within smaller timeframes, such as a month, a season or a year, and climate change considers changes that occur over a longer period of time, typically over decades or longer.

14.2 Factors Affecting Climate Change

A. Natural Factors

- ☆ **Continental Drift:** They have formed millions of years ago when the landmass began to drift apart due to plate displacement. This impacts climate change due to the change in the landmass's physical features and position and the change in water bodies' position like the change in the follow of ocean currents and winds.
- ☆ **Volcanism:** Volcanic eruption emits gasses and dust particles that last for a longer period causing a partial block of the Sun rays thus leading to cooling of weathers and influencing weather patterns.
- ☆ **Changes in Earth's Orbit:** A slight change in the Earth's orbit has an impact on the sunlight's seasonal distribution reaching earth's surface

across the world. There are three types of orbital variations, variations in Earth's eccentricity, variations in the tilt angle of the Earth's axis of rotation and precession of Earth's axis. These together can cause Milankovitch cycles, which have a huge impact on climate and are well-known for their connection to the glacial and interglacial periods.

B. Anthropogenic Factors

- ☆ **Greenhouse Gasses:** These absorb heat radiation from the sun resulting in an increase in Global Temperature. GHGs mostly do not absorb solar radiation but absorb most of the infrared emitted by the Earth's surface. Global warming begins with the greenhouse effect, which is caused by the interaction between incoming radiation from the sun and the atmosphere of Earth.
- ☆ **Atmospheric Aerosols:** These can scatter and absorb solar and infrared radiation. Solar radiation scatters and cools the planet whereas aerosols on absorbing solar radiation increase the temperature of the air instead of allowing the sunlight to be absorbed by the Earth's surface. Aerosols have a direct effect on climate change on absorption and reflection of solar radiation. Indirectly it can affect by modifying clouds formation and properties. It can even be transported thousands of kilometers away through winds and circulations in the atmosphere.
- ☆ **Shift in land-use pattern:** Most of the forests and land covers are replaced by agricultural cropping, land grazing or for Industrial or commercial usage. The clearing of forest cover increases solar energy absorption and the amount of moisture evaporated into the atmosphere

14.3 Greenhouse Effect

The greenhouse effect is the rise in temperature of the earth due to certain gases in the atmosphere. The gases are responsible for the greenhouse effect called as greenhouse gases (*e.g.* CO_2, N_2O, CH_4, CFCs).

Greenhouse gases: Many greenhouse gases occur naturally in the environment, such as water vapor, carbon dioxide, methane, nitrous oxide and ozone. Others such ash's, CFC, hydrofluorocarbons, perfluorocarbons and sulfur hexafluoride.

a) Carbon Dioxide (CO_2)

Carbon dioxide enters the atmosphere through the burning of fossil fuels (oil, natural gas and coal), solid waste, trees and wood products, chemical reactions (manufacture of cement) and removed from the atmosphere when it is absorbed by plants as part of the biological carbon cycle. For each 10 per cent rises in CO_2 concentration, the temperature increases by 3°C. It's present proportion in atmosphere is 390 ppm.

b) Nitrous Oxide (N_2O)

Nitrous oxide is emitted during various agricultural and industrial activities, as well as during combustion of fossil fuels and solid waste. The proportion of nitrous oxide in air is 0.25-0.35 ppm.

c) Methane (CH_4)

Methane is emitted during the production and transport of coal, natural gas, oil and organic waste decomposes. The proportion of methane in air is about 1.3 to 1.6 ppm.

d) Fluorinated Gases

CFC, HFC, PFC and sulfur hexafluoride are synthetic, powerful greenhouse gases that are emitted from a variety of industrial processes. Fluorinated gases are sometimes used as substitutes for ozone-depleting substances. They are referred as high global warming potential gases (high GWP gases). CFC is mainly responsible for ozone depletion in the atmosphere.

Sunlight enters the earth's atmosphere, passing through the greenhouse gases. As it reaches the earth's surface, land, water and biosphere absorb the sunlight's energy. Once absorbed, this energy is sent back into the atmosphere. Some of the energy passes back into space, but much of it remains trapped in the atmosphere by the greenhouse gases. This is the completely natural process and without these gases all the heat would escape back into space. The greenhouse effect is very important process, because without the greenhouse effect, the earth would not be warm enough for humans to live. But if the greenhouse effect becomes stronger, it could make the earth warmer than usual.

14.4 Global Warming

Global warming is the rise of the average temperature of the earth's climate system. The solar energy that passes through the earth's atmosphere is absorbed by the surface of the earth while a major portion of it is reflected into space. Greenhouse gases like carbon dioxide, nitrous oxide, methane, create a partial blanket over the earth's atmosphere and do not allow the outgoing infra-red radiations to travel back into space. They trap the infra-red radiations and reflect them into the earth's atmosphere. Such effect is known as the 'Greenhouse Effect'. As a consequence of the greenhouse effect, the global mean temperature of the earth increases. Such increase in temperature continues over a long period of time and is the main cause of extensive warming of the planet is known as global warming. Industrialization, land use change, overpopulation, deforestation and change in lifestyles are responsible for global warming.

14.5 Potential Effects of Climate Change in India

- ☆ **Extreme Heat:** India is already experiencing a warming climate. Unusual and unprecedented spells of hot weather are expected to occur far more frequently and cover much larger areas. Under 4°C warming, the west coast and southern India are projected to shift to new, high-temperature climatic regimes with significant impacts on agriculture.

- **Changing Rainfall Patterns:** A decline in monsoon rainfall since the 1950s has already been observed. A 2°C rise in the world's average temperatures will make India's summer monsoon highly unpredictable. At 4°C warming, an extremely wet monsoon that currently has a chance of occurring only once in 100 years is projected to occur every 10 years by the end of the century.
- **Droughts**: Evidence indicates that parts of South Asia have become drier since the 1970s with an increase in the number of droughts. In 1987 and 2002-2003, droughts affected more than half of India's crop area and led to a huge fall in crop production. Droughts are expected to be more frequent in some areas, especially in north-western India, Jharkhand, Orissa, and Chhattisgarh. Crop yields are expected to fall significantly because of extreme heat by the 2040s.
- **Groundwater**: Even without climate change, 15 per cent of India's groundwater resources are overexploited. Falling water tables can be expected to reduce further on account of increasing demand for water from a growing population, more affluent lifestyles, as well as from the services sector and industry.
- **Glacier Melt:** Most Himalayan glaciers have been retreating over the past century. At 2.5°C warming, melting glaciers and the loss of snow cover over the Himalayas are expected to threaten the stability and reliability of northern India's primarily glacier-fed rivers. Alterations in the flows of the Indus, Ganges, and Brahmaputra rivers could significantly impact irrigation, affecting the amount of food that can be produced in their basins as well as the livelihoods of millions of people.
- **Sea level rise:** With India close to the equator, the sub-continent would see much higher rises in sea levels than higher latitudes. Sea-level rise and storm surges would lead to saltwater intrusion in the coastal areas, impacting agriculture, degrading groundwater quality, contaminating drinking water, and possibly causing a rise in diarrhea cases and cholera outbreaks, as the cholera bacterium survives longer in saline water. **Cities near the seas** are vulnerable to the impacts of sea-level rise,tropical cyclones and riverine flooding.

14.6 India's Response to Climate Change

- **National Action Plan on Climate Change** (NAPCC): outlines existing and future policies and programs addressing climate mitigation and adaptation. The Action Plan identifies eight core "national missions" running through to 2017: Solar Energy; Enhanced Energy Efficiency;

Sustainable Habitat; Water; Sustaining the Himalayan Ecosystem; Green India; Sustainable Agriculture; and Strategic Knowledge for Climate Change. Most of these missions have strong adaptation imperatives.

- **National Clean Energy Fund**: The Government of India created the National Clean Energy Fund (NCEF) in 2010 for financing and promoting clean energy initiatives and funding research in the area of clean energy in the country. The corpus of the fund is built by levying a cess of INR 50 (subsequently increased to INR 100 in 2014) per tonne of coal produced domestically or imported.
- **Paris Agreement**: Under the Paris Agreement, India has made three commitments. India's greenhouse gas emission intensity of its GDP will be reduced by 33-35 per cent below 2005 levels by 2030. Alongside, 40 per cent of India's power capacity would be based on non-fossil fuel sources. At the same time, India will create an additional 'carbon sink' of 2.5 to 3 billion tonnes of Co2 equivalent through additional forest and tree cover by 2030.
- **International Solar Alliance**: ISA was launched at the United Nations Climate Change Conference in Paris on 30 November 2015 by India and France, in the presence of Mr. Ban Ki Moon, former Secretary-General of the United Nations.
- **Bharat Stage Emission Norms**: Emissions from vehicles are one of the top contributors to air pollution, which led the government at the time to introduce the BS 2000 (Bharat Stage 1) vehicle emission norms from April 2000, followed by BS-II in 2005. BS-III was implemented nationwide in 2010. However, in 2016, the government decided to meet the global best practices and leapfrog to BS-VI norms by skipping BS V altogether.

Chapter 15

Modification of Microclimate

15.1 Introduction

Micrometeorology is a branch of meteorology which deals with atmospheric phenomenon over a few square kilometer areas and within a few meters of height above the ground.

The climate of a small area or climate near ground surface to the top of the plant is known as micro-climate. It is also called as the plant climate or ecoclimate or habitat climate. Macro-climate is the climate of a large area. The climate of small area intermediate between microclimate and microclimate such as land-sea breeze and mountain-valley breeze is known as meso-climate. Eco-climate is a microclimate just above a crop and within the canopy. Phyto-climate is modified form of microclimate produced due to influence of crops. It is controlled by the structure of plant cover, plant height, number of leaves, tillers, leaf area, distribution of plants and spacing between plants and their orientation towards sunlight.

Crop microclimate is the climate just above and within the crop canopy and in the soil root zone that can be influenced by day-to-day management practices at varies time scales. It is any climate condition that exists within a few meters or less above and below the earth's surface and within vegetation canopies. The best crop microclimate is one that provides the most favorable environment for the desired plant response which maximizes crop productivity.

Microclimatic modification is an artificial control of field environment to keep the optimum condition of plant growth and crop production. It is an attempt to regulate the elements of climate on a micro scale resulting in a climate that is favorable for plant growth. It is any process in which all the adverse or stress conditions prevailing in the immediate vicinity of the plants are modified for the better crop growth, development and yield.

15.2 Processes of Microclimate Modification

A) Controlling the Heat Load

Heat load on crop canopy can be increased called heat trapping, or decreased called heat evasion.

i) Heat Evasion

It can be achieved by shading the plants, irrigation management or by using antitranspirants. Shading of plants is a common method of evading of solar radiation. A number of shade structure are used and they are opaque. The material wood or fiber act as a thermostat and can be applied or removed as required. It keeps temperature low and retards evapotranspiration.

ii) Heat Trapping

It can be achieved by adopting appropriate row direction or plant density. Planting trees on sunny slope and erecting alternate rows of low stone walls is useful techniques in heat trapping. The low stone walls reflect the light back towards the lower portion and shaded sides of the trees. Heat sensitive crops can be covered with plastic enclosure during night.

B) Controlling Water Balance

It involves increasing the amount of water stored in the root zone (by strip cropping, contour cropping, ploughing, terracing or bunding), increasing infiltration (by increasing tillage or increasing row spacing), reducing soil evaporation (by using mulches and crop residues, plastic covers or thick cultivation crops), modifying transpiration (by using antitranspirants, substances forming film on leaf surface, stomata closing materials and reflectants).

i) Mulching

It is the process of covering soil with a material to improve the condition of the soil. Mulch is any material applied on soil surface to check evaporation, suppress weeds, reduce water loss and improve soil moisture. Mulch is used to provide the benefits like reduce weeds by blocking sunlight, reducing water requirements by preventing evaporation, keep the soil cooler in summer and warmer in winter, reduce soil moisture losses *etc.*

Types of Mulches

- ☆ **Soil mulch or dust mulch***:* If the soil surface is loosened, it acts as a mulch to reduce evaporation. It is called soil mulch or dust mulch. Intercultivation creates soil mulch in growing crops. Its usefulness is doubtful in alfisols but helps in closing deep cracks in vertisols.

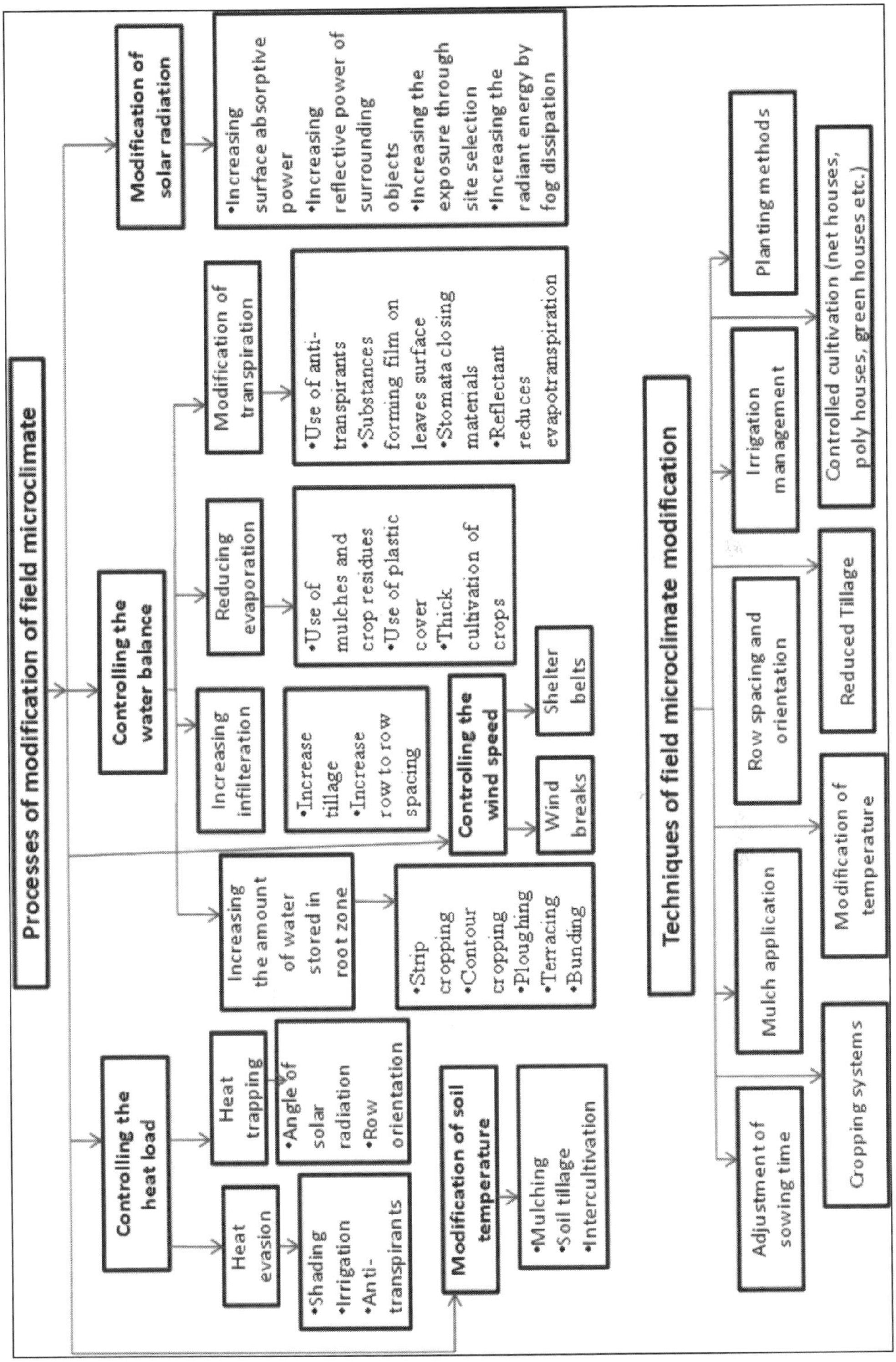
Processes of modification of field microclimate
Controlling the heat load
Heat evasion
•Shading •Irrigation •Anti-transpirants
Heat trapping
•Angle of solar radiation •Row orientation
Modification of soil temperature
•Mulching •Soil tillage •Intercultivation
Controlling the water balance
Increasing the amount of water stored in root zone
•Strip cropping •Contour cropping •Ploughing •Terracing •Bunding
Increasing infilteration
•Increase tillage •Increase row to row spacing
Reducing evaporation
•Use of mulches and crop residues •Use of plastic cover •Thick cultivation of crops
Modification of transpiration
•Use of anti-transpirants •Substances forming film on leaves surface •Stomata closing materials •Reflectant reduces evapotranspiration
Controlling the wind speed
Wind breaks
Shelter belts
Modification of solar radiation
•Increasing surface absorptive power •Increasing reflective power of surrounding objects •Increasing the exposure through site selection •Increasing the radiant energy by fog dissipation
Techniques of field microclimate modification
Adjustment of sowing time
Mulch application
Row spacing and orientation
Irrigation management
Planting methods
Cropping systems
Modification of temperature
Reduced Tillage
Controlled cultivation (net houses, poly houses, green houses etc.)

- ☆ **Stubble mulch:** Crop residues; wheat straw/cotton stalks are left on the soil surface as a stubble mulch.
- ☆ **Stone mulch:** Mainly useful in rainfed agriculture system of the arid region. Stones/pebbles are used as covering soil.
- ☆ **Plastic mulch:** Plastic materials like polyethylene, polyvinyl chloride is used as mulching materials. Non degradable and Degradable are two types of plastic mulch. Degradable plastic mulch are; a) Biodegradable type plastic films are converted through microbial activity in soil to carbon dioxide, water and natural substance. This type of plastic mulch film is easily degraded in the soil over a period of time. b) Photo degrable type of plastic mulch film gets destroyed by sun light in a shorter period.

Selection of Mulches for Mulching

Sl.No.	*Condition*	*Types of Mulch's*
1	Rainy season	Perforated mulch
2	Orchard and plantation	Thicker mulch
3	Soil solarization	Thin transparent film
4	Weed control through solarization	Transparent film
5	Weed control in cropped land	Black film
6	Saline water use	Black film
7	Sandy soil	Black film
8	Summer cropped land	White film
9	Insect repellent	Silver colour film
10	Early germination	Thinner film

- ☆ **Vertical mulching:** To prolong the beneficial effect of subsoiling, a method called vertical mulching has been developed. The object of vertical mulching is to fill slots with organic matter and keeping them open and functional for a longer period. In black soils, to improve infiltration and storage of rainwater vertical mulches are formed. It consists of digging narrow trenches across the slope at intervals and placing the straw or crop residues in these trenches. *e.g.* The pruned plant material is placed in contour trenches formed between rows of coffee or in trenches around the plants in concentric each year in one circle.

ii) Antitranspirant

Antitranspirants are substances applied to the plants for the purpose of reducing transpiration (water loss) without causing a significant effect on other plant processes, such as photosynthesis and growth.

Types of Antitranspirant

- ☆ Stomatal closing: They induce stomatal closing or decrease size and number of stomata which subsequently reduce the photosynthesis. *e.g.* phenyl mercuric acetate.
- ☆ Film forming: Plastic and waxy material which form a thin colourless film over the leaf surface and result in a physical barrier. These glossy films then reduce water loss on plants. *e.g.* silicone oils, waxes.
- ☆ Reflecting: These are increasing the reflection of light from leaf surface thus reducing leaf heating and water losses. *e.g.* Kaoline.
- ☆ Growth retardant: These chemicals reduce shoot growth and increase root growth and thus enable the plants to resist drought. *e.g.* cycocel.

C) Controlling Winds

Winds can be controlled by applying wind breaks or shelter belts. These protections reduce wind speed on the leeward side and save the plants from freezing injury and mechanical damage.

i) Windbreaks

A windbreak is any barrier (natural or artificial) that reduces troublesome winds by creating a wind shadow to the leeward side. A windbreak protects a distance from 2 to 10 times its height. So, a 5 meter high windbreak would protect a distance up to 50 meters away. Windbreaks are most effective when oriented at right angles to prevailing or troublesome winds. Windbreak density is the ratio of the solid portion of the barrier to the total area of the barrier.

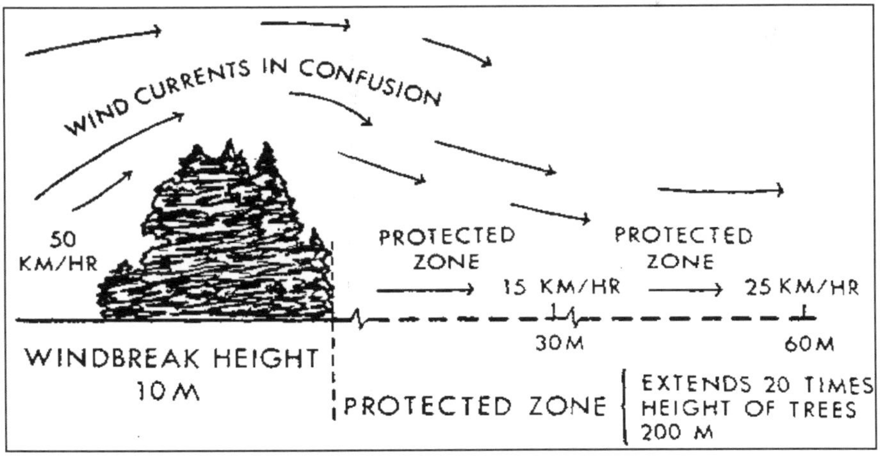

Figure 15.1. Windbreaks.

Wind flows through the open portions of a windbreak the more solid the windbreak, the less wind passes through. For best protection the length of a windbreak should exceed the height by 10:1 ratio.

Purpose of raising wind breaks is;

- ☆ To protect field crops or livestock from cold and hot wind.
- ☆ To prevent soil erosion.
- ☆ To reduce evaporation from farmlands.
- ☆ To improve the microclimate.
- ☆ For fencing and boundary demarcation.
- ☆ For productive role-fuel, fodder.

ii) Shelter Belts

Shelter Belts are the belts or blocks consisting of several rows of trees. It should be designed to contain ten or more rows of tree and shrubs to provide maximum benefits to wildlife and homeowners. Placement is primary on the north and west sides of farmsteads to block the prevailing winds. Four important design factors; snow catch, height of the lift trees, number of evergreen rows and overall width of the belt from west to east and north to south will determine the effectiveness of any shelterbelt. Shrubs are planted in the outermost rows to catch drifting snow. In middle, four rows of tall deciduous trees lift chilling winds above the farmstead.

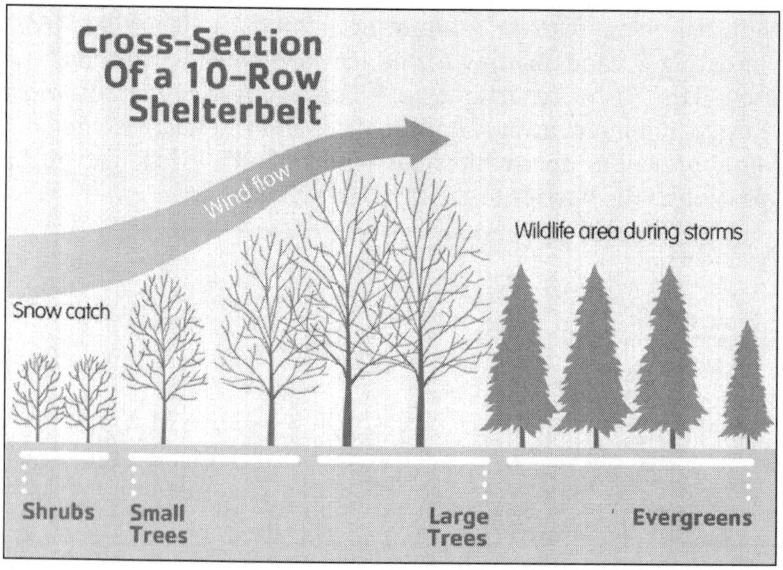

Figure 15.2. Shelter Belts.

Purpose of raising shelter belts is;

- ☆ To deflect air currents.
- ☆ To reduce the velocity of winds.
- ☆ To provide general protection to the leeward areas against the effects of wind erosion.

- ☆ To protect the leeward areas from desiccating effects of hot wind.
- ☆ To provide fuel, fodder timber.

Modification of Solar Radiation

It can be accomplished by increasing the surface absorptive power, reflective power of the surrounding objects, exposure through site selection, increasing the radiant energy by fog dissipation and by adopting appropriate row direction.

Soil temperature can be modified by mulch application and soil tillage *etc.* Mulch helps in reducing the soil temperature in summer and increasing during winter season. Similarly, soil tillage by harrowing or inter-cultivation also helps in modifying the soil thermal regime.

Protected cultivation under controlled environment like greenhouses (for temperature, humidity and CO_2 control), polyhouses (naturally ventilated or with controlled environment) and shade houses (for crop production in warm climate) provide other means for raising crops under controlled microclimatic conditions.

Chapter 16

Climatic Normals for Crops

16.1 Rice

Rice is a heat loving crop and short-day plant. It requires hot and humid climate, high humidity, prolonged sunshine and an assured supply of water. It can be cultivated from sea level to as high as 3000 meters. Total growing period ranges 90-150 days for rice cultivation. The average temperature required throughout the life period of the crop ranges is 21-37°C. The optimum temperature range for germination is 22-31°C, for flowering is 22 to 23°C and for ripening is 20-25°C. At the time of tillering, the crop requires a higher temperature than for growth. The base or lower threshold temperature is **10°C**. If the temperature is high at grain filling stages, the respiration rate is accelerated and grain filling period is reduced. The rice yield is greatly reduced by formation of sterile spikelet's, which are induced by low temperatures. On an average the critical low temperature for inducing sterility is 15-17°C. It requires abundant moisture. Crop is generally grown in regions with rainfall about 1250 mm. Its seasonal water requirement ranges 750-2500 mm. The feet of the plant should remain submerged in water from time of sprouting to the milk stage of the grain.

16.2 Wheat

Wheat is cold loving crop and long day plant. It can be grown not only in the tropical and sub-tropical zones but also in the temperate zone. It requires cool, moist weather during the growing period followed by dry, warm weather to enable the grain to ripe properly. It can be cultivated from sea level to as high as 3300 meters. Total growing period ranges 115-120 days for wheat cultivation.

The optimum temperature requirement for growth and development of wheat is 15-20°C. The base temperature is 5°C. Higher temperature 30-35°C have in general detrimental effects on the wheat crop *i.e.* reduce in tillering. Cloudy weather with high humidity and low temperatures are conducive for rust attack. Temperatures above 25°C during grain filling period tend to depress grain weight. Wheat is grown in regions with rainfall 500-875 mm. Its moisture requirements range 450-650 mm. Soil moisture stress at any stage decreases grain yield.

16.3 Cotton

Cotton is a heat loving crop and short-day plant. It requires abundant sunshine, adequate soil moisture and fairly high temperature. Cotton plant cannot stand frost and hence its cultivation is restricted to an altitude of 1000 meters only. The growing period ranges 180 to 240 days for cotton cultivation. The maximum and minimum soil temperature limits are 14°C and 40°C. The minimum, optimum and maximum cardinal temperatures for cotton growth is 15-20°C, 25-30°C and 35-40°C, respectively. The optimum temperature required for germination is 16°C, for growth is 15-20°C, for proper vegetative growth is 21-22°C, for fruiting is 26-32°C and for boll and Fibre maturation is 22-27°C. Low temperatures of 15-20°C retard growth and reduce fruiting. High temperatures 30-40°C and above have detrimental effects on growth and fiber quality. Cloudiness prolongs vegetative growth. Abundant sunshine during the period of boll maturation and harvesting is essential to obtain a good quality produce. Cotton is grown in regions having 550-750 mm rainfall. Its water requirement ranges 700-1300 mm. Continuous rains during flowering cause improper pollination and bad quality fiber in boll developed.

16.4 Sugarcane

Sugarcane is a tropical plant but it can grow in sub-tropics also. The minimum soil temperature for the germination of sugarcane setts is in the range of 19-21°C. The optimum temperature range is from 27-38°C, while temperatures above 38°C are not conducive. No root development occurs below 12°C. The minimum threshold temperature for cane growth is 16°C and tillering is maximum at 30°C. Maximum temperature above 35°C and minimum temperature less than 18°C during tillering phase decrease cane yield. Optimum day and night temperatures for elongation are 31°C and 20°C, respectively. Maximum temperature above 37°C inhibits growth and growth also stops below 7°C. Growth and yield both fails with low temperature 0°C during winter nights. During ripening cool (10-20°C), dry sunny weather is required. Longer nights and shorter days, bright sunshine, and low humidity are conductive to effective dehydration and rapid accumulation of sucrose. The yield of the crop is reduced to one half if sunshine is reduced to half of its normal value. Sugarcane crop grown in area having 125-165 cm rainfall. Its water requirement ranges 1500-2500 mm. Its sensitivity to water supply is high. For harvesting and handling it requires comparatively dry conditions. Deficient rainfall produces a fibrous cane while heavy rain results in low sugar content in cane. The tillering is increase in bright sunshine with larger day lengths and hot weather favors tillering. Cold weather arrest growth, in' hot weather borer attack is increased. Cool and dry climate with less than 50 per cent relative humidity is beneficial for ripening of cane.

16.5 Onion

The onion is cool season and photosensitive crop, tolerant to frost in the young stage but less sensitive to heat. It is well adapted to a temperature range of 13-25°C. A temperature range of 15-21°C before bulbing is required for its good vegetative growth whereas a temperature of 20-25°C is considered ideal for bulb development. Very low temperatures in the beginning favour physiological disorder called bolting. Sudden rise in temperature favours early maturity of the crop in rabi and results in small size of bulbs. At least 10 hours light with favorable temperature is essential for growth, development and bulbing. In identical light quality, photoperiod and temperature, higher the light intensity more rapid bulbing and earlier ripening of bulb resulted. Optimum soil temperature for germination, vegetative growth, bulb development is 15-25°C, 13-23°C and 20-25°C respectively. It requires about 70 per cent relative humidity for good growth. Crop thrives well in places, which receive an average rainfall of 750-1000mm during monsoon.

16.6 Maize

Maize is a warm weather crop and is not grown in areas where the mean daily temperature is less than 19°C or where the mean of the summer months is less than 23°C. Although the minimum temperature for germination is 10°C, germination will be faster and less variable at soil temperatures of 16 to 18°C. At 20°C, maize should emerge within five to six days. The critical temperature detrimentally affecting yield is approximately 32°C. Frost can damage maize at all growth stages and a frost-free period of 120 to 140 days is required to prevent damage. Ovule number increases with increase in temperature. While the growth point is below the soil surface, new leaves will form and frost damage will not be too serious. Leaves of mature plants are easily damaged by frost and grain filling can be adversely affected. Water Approximately 10 to 16 kg of grain are produced for every millimeter of water used hence, rain requires between 350 and 450 mm of per annum.

16.7 Potato

It is a temperate or cool season crop, which needs low temperature, low humidity and bright sunny days. The vegetative growth of the plant is best at a temperature of 24°C, germination at 25°C and tuber development is favoured at 20°C. The colder the soil temperature, the more rapid the initiation of tubers and the greater the number of tubers formed. The optimum soil temperature for tuber initiation is 15 to 20°C. The potato tuber is an enlarged portion of the stolon. The initiation of this tuber is triggered by short day lengths and involves growth hormones. Longer day lengths delay tuber initiation and favor the growth of the stolon and shoot. High temperatures also reduce tuber formation. Temperature above 30°C stop tuber formation because at higher temperature rate of respiration increase. Cloudy day, rains and high humidity are very congenial for spread of diseases.

16.8 Tobacco

It is a tropical crop in origin but it is grown in subtropical and temperate regions of the world. The crop needs around 28°C as an optimum temperature for germination but it may germinate at a temperature ranging between 15°C to 35°C. The crop cannot withstand frost and starts withering or wilting at a temperature above 35°C. Tobacco needs 50-100 cm rainfall which should be well distributed throughout growing period. Areas having a rainfall of more than 125 cm are not fit for tobacco production, whereas a rainfall of less than 75 cm results in production of leathery leaves having very high nicotine content in them. The leaves remain smaller and possess very poor burning quality. Rainfall at ripening stage will wash the gums from the leave's surface resulting in poor curing. Water logging at any stage of growth is deadly harmful because the plants cannot withstand water logging. While hailstorm causes a serious injury to the leaves any time during growing period. Relative humidity (85-90 per cent) is an optimum humidity for the crop because at this level of humidity the leaves are easily used and they do not become very brittle. Wind storm at any stage of growth after seedling establishment is very harmful because the leaves split and reduce the quality specially when the plants have a large exposed leaf surface. During cold weather and at low temperature there is slow growth of the plants and this also affects the quality of the crop.

16.9 Soybean

Soybean is a tropical crop but its cultivation now extended to subtropical and temperate regions or climates. Climatic requirements of soybean are almost similar to that of maize. Minimum and maximum temperature for seed germination is 5 and 40°C respectively. Minimum temperature for growth is 10°C and maximum rate of development between planting and flowering occurs at 30°C. High night Temperature increases vegetative growth and no of branches. Low temperature delay flowering. Flowering is poor below 20°C and increases up to 32°C beyond which it increases flower abscission and pod setting. The yield reduces to half above 33°C. Soil temperature of about 30°C increases seed yield by more than 50 per cent while 10°C decrease the yield by similar magnitude. Soybean is basically a short-day plant. Day length influences the rate of development. In short day types, increased day length may result in the delay of flowering and taller plants with more nodes. Soybean is often cultivated as kharif rainfed crop and as rabi crop on receding soil moisture after kharif crop. It can come up in areas with rainfall varying from 600 to 1000 mm. Soil moisture stress at flowering and pod development reduces the seed yield by about 50 per cent. Excess soil moisture inhibits nitrogen fixation. Low atmospheric humidity significantly reduces the seed yield.

16.10 Chick Pea

Chickpea is a winter legume withstanding temperature between 8°C minimum and 22°C maximum during the coldest month. The growing period ranges 90 to 150 days for chick pea cultivation. During the flowering period the average minimum temperatures 10-14°C and average maximum temperature between 25-31°C are considered optimum for hybridization and seed setting in most cultivars. Chickpea is sensitive to both light and temperatures. It requires long days and cool temperatures. Average around 20°C with warm days (20°-25°C) and cool (5°-10°C) or even cold frosty nights (-1° to 0°C). The optimum soil temperature for establishment of desi chickpea between 25-30°C. Chick pea crop is grown in area having 600-1000 mm rainfall. Its water requirement ranges 100-450 mm.

16.11 Pearl Millet

Millets are heat loving plants. It is mostly distributed in sub-humid and semi-arid climates. For germination, the minimum temperature required is 8-10°C. For vegetative growth, the optimum temperature required is 28-32°C. The plants are not frost-resistant; when temperatures fall below 1°C, the young leaves of plant are damaged and growth is retarded. A temperature of 35 to 40°C is normally tolerated by them. Millets have low moisture requirements and are highly drought-resistant. Pearl millet crop grown in area having 400-600 mm rainfall. Rainfall during flowering and grain formation cause poor grain setting. Rain, high humidity and low temperature at grain maturity cause ergot disease. Higher temp at this stage induces early flowering.

16.12 Sunflower

It is short duration, photo insensitive, wide adaptability and drought tolerance crop. It can be grown anytime of the year and can serve an ideal catch crop during the period well and is kept follow. It tolerates temperature range of 5°C to 40°C without significant yield reduction indicating that it is adapted to the region with warm days and cool nights. However, it grows well within a temperature range of 20-25°C. The optimum temperature being 27 to 28°C. High humidity accompanied with cloudy weather and rainfall at the time of flowering results in poor set. The amount of linoleic acid content decrease with high temperature at maturity. It can grow in the areas receiving rainfall in the range of 200 to 750 mm. It requires cool climate during germination to seedling growth, warm weather during seedling stage to flowering stage and warm and sunny days during flowering to maturity stage.

16.13 Sorghum

Sorghum is a tropical crop, drought resistant crops and hence ideal crop for cultivation in a region which depends entirely upon rainfall. It thrives well in a region having min temperature of over 27°C. The average temperature during the growing season should be 21 to 32°C. It regulates transpiration and remains dormant during the period of drought and starts active growth when the moisture status of the soil improves. It can grow in areas receiving 350 to 1000 mm rainfall. Excessive soil moisture as well as prolonged drought is harmful for the crop growth.

Chapter 17

Advance Agricultural Meteorology

17.1 Remote Sensing

It is the art and science of obtaining information from a distance, *i.e.* obtaining information about objects or phenomena without being in physical contact with them. Pisharoth Rama Pisharoty is considered to be the father of remote sensing in India.

17.1.1 Process of Remote Sensing

Sun is the principal source of energy. When the energy (in the form of electromagnetic radiation) reaches the earth's atmosphere, it undergoes the process of reflection, absorption and transmission. Earth's surface consists of different natural and man-made features which reflect, absorb, store and emit earth's radiation at different wavelengths in different percentages, depending upon their physical and chemical properties. Remote sensing sensors record different amount of radiation that is reflected or emitted from different earth surface features and reproduce it in form of an image.

17.1.2 Application of Remote Sensing

- Monitoring droughts and floods
- Crop production forecasting
- Water resource mapping
- Soil mapping
- Land mapping

- Climate change monitoring
- Collection of weather data
- Identification of problematic soils
- Assessment of crop damage and crop progress
- Identification of pests and disease infestation
- Soil moisture estimation
- Crop intensification

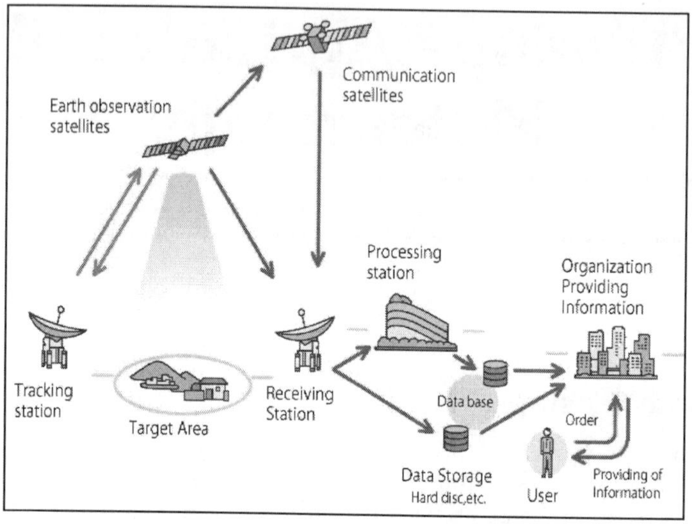

Figure 17.1. Remote Sensing.

17.2 Crop Modelling

Model is an equation or set of equations which represent the behavior of a system. Simulation is the process of building models and analyzing the system. Crop model is the simple representation of a crop. Crop simulation model (CSM) is a simulation model that describes processes of crop growth and development as a function of weather conditions, soil conditions and crop management.

17.2.1 Types of Models

Statistical Model

These models express the relationship between yield or yield components and weather parameters. In these models' relationships are measured in a system using statistical techniques. The model does not explain the mechanism, approach and aspect of weather parameters by which they influence yield. The model does not take soil variability, genetic potential and management into account.

Mechanistic Model

These models explain not only the relationship between weather parameters and yield, but also explain the mechanism of influencing dependant variables. These models are based on physical selection.

Deterministic Model

These models estimate the exact value of the yield or dependent variables. These models also have defined coefficients.

Stochastic Model

These models use the value of weather parameters at some or other probability level. Hence, output components are estimated within a range depending upon probability level of dependent variables. These models define or state of dependent variable at a given rate.

Dynamic Model

Time is included as a variable. Both dependent and independent variable are having value which remain constant over a given period of time.

Static Model

Time is not included as a variable. Both dependent and independent variable are having value which remain constant over a given period of time.

Descriptive Model

These models are helpful to determine quickly the weight of the crop where no observation was made.

Explanatory Model

These models consist of quantitative description of the mechanisms and process that cause the behavior of the system. It contains description of distinct processes such as leaf area expansion and tiller production.

Simulation Model

These models use one or more sets of different equation and calculate both rate and state variables over time. It is used to estimate agricultural production as a function of weather, soil condition and crop management.

17.2.2 Uses of Crop Simulation Model

- ☆ Yield forecasting
- ☆ Assessment of climate variability and climate change
- ☆ Site specific farming
- ☆ Introduction of a new crop
- ☆ Study of insect and disease development

- Yield gap analysis
- Evaluation of potential yield
- Genetic improvement
- Plant breeding

17.2.3 Examples of Crop Simulation Model

Model	*Description*
DSSAT	Decision Support System for Agrotechnology Transfer
APSIM	Agricultural Systems Modelling and Simulation
Aquacrop	It is developed by FAO
EPIC	Environmental Policy Integrated Climate model
CropSyst	Cropping Systems Simulation Model
STICS	Simulateur multidisciplinary pour les Cultures Standard
SUCROS	Simple and Universal Crop growth Simulator
WOFOST	World Food Studies
DNDC	Denitrification-Decomposition model
DAISY	Danish Simulation Model
CERES	Crop Environment Resource Synthesis

17.3 GIS (Geographic Information System)

It is a system for capturing, storing, manipulating, analyzing and displaying data which are spatially referenced to the earth. This system comprises hardware, software, database, procedures and expertise designed to support the storage, manipulation, analysis and display data of feature attributes and location data to produce maps. From these, information on vegetation, soil types, plant stress, crop infestation, information on agricultural land use planning are collected for the purpose of timely decision making.

17.4 GPS

Global positioning system is a satellite-based radio-navigation systems that provides geo-location and time information to a GPS receiver anywhere on or near the earth that helps users to record positional information with an accuracy between 100 to 0.01 m. This system comprises ground control segments, space segment and user segment to provides geo-location.

Applications of GPS

- Location: determining a basic position.
- Navigation: getting from one location to another.

- ☆ Tracking: monitoring the movement of people, animals and goods.
- ☆ Mapping: creating maps of the world.

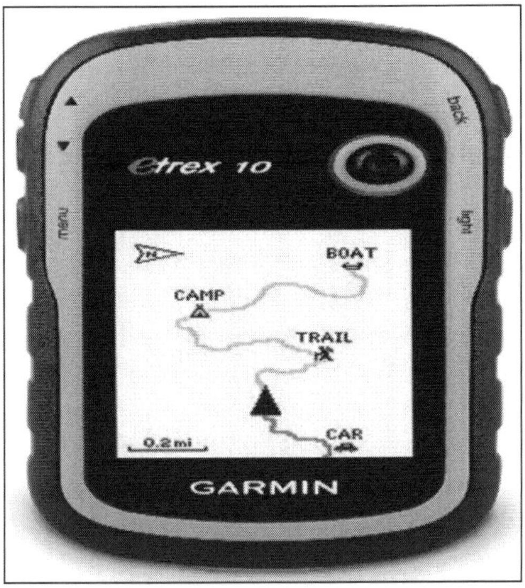

Figure 17.2. Global Positioning System.

17.4 AWS

An Automatic Weather Station (AWS) is a facility that automatically transmits or records observations obtained from measuring instruments. In an AWS, the measurements of meteorological elements are converted into electrical signals through sensors. The signals are then processed and transformed into meteorological data. The information is finally transmitted the by wire or radio or automatically stored it on a recording medium. The parts of AWS can be summarized as;

Data logger field unit: This unit collects and stores the weather data in the field. The data logger collects information from every sensor (temperature, RH, wind speed, wind direction, rainfall, solar radiation, *etc.*) and stores it either in its own memory or a memory card. It also processes most of the meteorological data *i.e.* provides average, minimum and maximum values of the recorded parameters.

Data logger terminal: The instructions to the field unit are given through the terminal, in other words, the terminal instructs the field unit about which sensor to use, what channel these sensors are wired to, when to store data, and how to label and organize data. It also helps in viewing the data on the display, check battery life and remaining memory on the data storage pack.

Data storage pack: The tabulated data is stored in the practical units of scientific measure using the data storage pack. For unloading the stored data, the data storage module can be connected to the data logger field unit. This unloaded data can then be transferred to a personal computer and stored on the computer hard disk for permanent record and further analysis.

Solar panels: Solar panels provide the power to run the weather stations. The solar panels are equipped with rechargeable lead-acid batteries.

Temperature sensor: The temperature probe is used to measure air temperature. Temperature is sensed by thermistor which is extremely sensitive and exhibits a large resistance change with small change in temperature. Soil temperature is measured by another temperature probe which is electrically identical to air temperature probe, but is physically more rugged for burial applications.

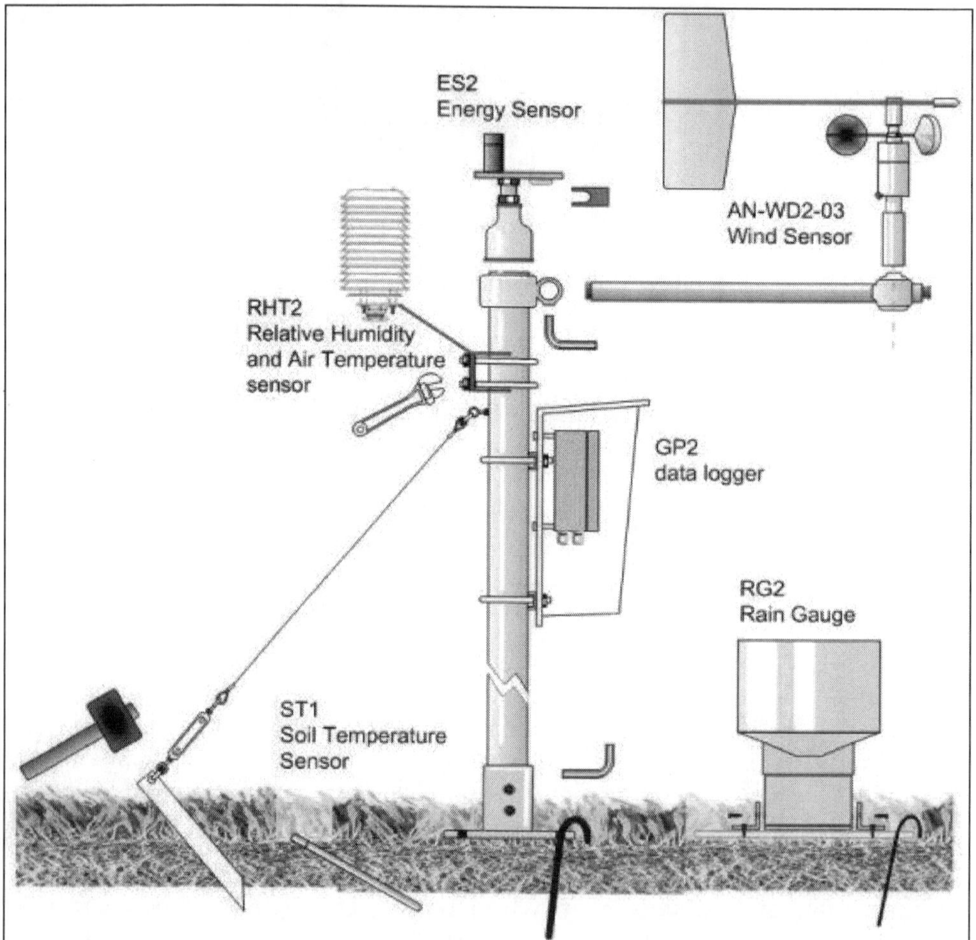

Figure 17.3. Automatic Weather Station.

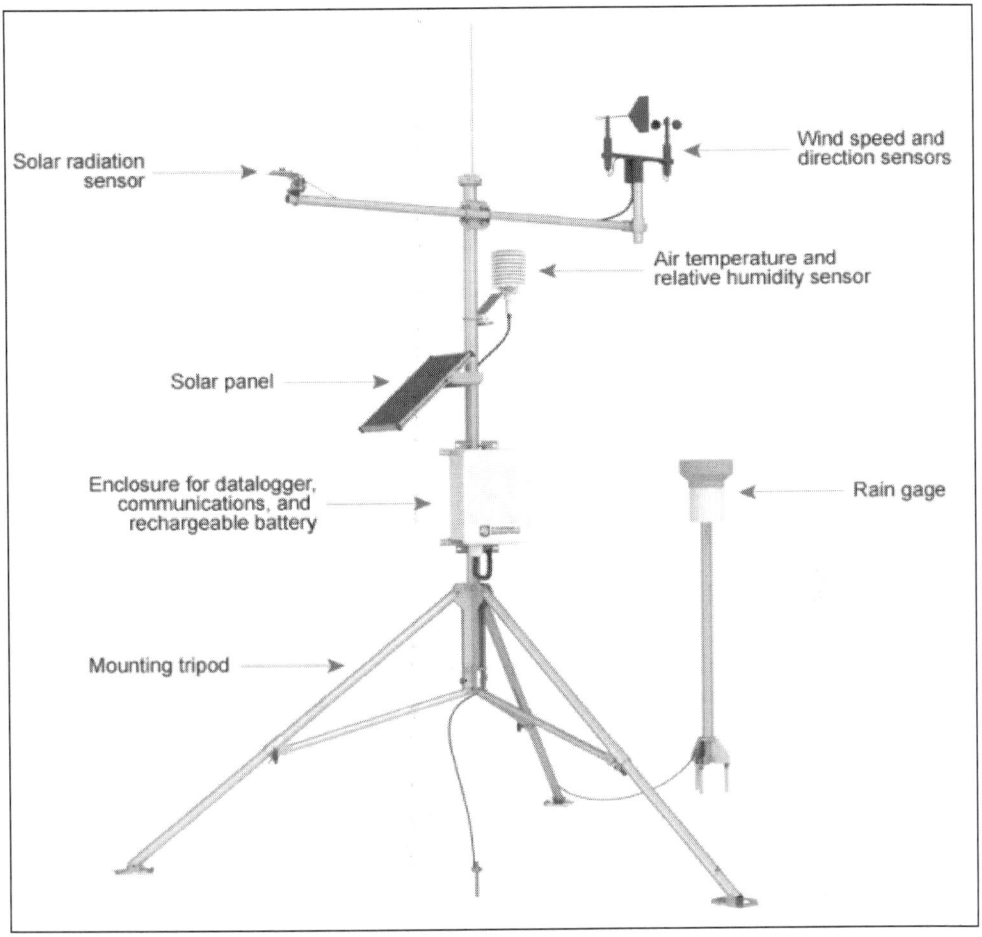

Figure 17.3. Automatic Weather Station.

Humidity sensor: Relative humidity and temperature probe sensors have ±1 per cent and ±0.3°C accuracy at 20°C. Measurement range for RH sensor is 0-100 per cent.

Wind direction sensor: The wind vane measures wind direction from 0° to 360° with a 5° accuracy. The sensor utilizes a potentiometer to vary the sensor resistance in relation to wind direction.

Wind speed sensor: The anemometer measures wind speed in the range of 0-45 ms^{-1} (0–160 kmh^{-1}). This sensor is a three-cup wheel assembly utilizing a magnet activated reed switch whose frequency is proportional to wind speed.

Radiation sensor: This sensor is designed for field measurement of sun and sky radiation. The silicon pyranometer puts out a current which is dependent upon the solar radiation incident upon the sensor. The current is measured as the voltage drop across a fixed resistor.

Rainfall sensor: This is a small adaptation of the standard Weather Bureau tipping bucket rain gauge. It measures rainfall at rates 50 mm per hour with an accuracy of ±1 per cent. It is designed such that one alternate tip of the bucket occurs for each 0.25 mm of rainfall. Each tip actuates a magnetic switch. The rain gauge should be mounted on a level ground and at least 30 cm above the ground surface.

The advantages of using AWS can be summarized as:

- Continuous observation is possible.
- Fully automated systems can also be installed at inaccessible sites.
- Observational can be obtained even when no staff are present.
- Reduces observer numbers and operating costs.
- Since meteorological data are taken as electrical signals, observer errors in reading are eliminated.
- Standardized observation techniques enable the homogenization of observed data in regions where automatic weather observation is adopted.
- New observation elements can be added relatively easily by installing new instruments.
- Optimal measuring instruments with the appropriate level of measurement accuracy for the required observation can be chosen, and the need for observer training is eliminated.

Chapter 18

Agro-Climatic Zones

18.1 Introduction

Food and Agriculture Organization (FAO) defined an agro-climatic zone as a land unit represented major climate and growing period, which is climatically suitable for certain range of crops and cultivars. It is a land unit in terms of its major climate, superimposed on the length of growing period or moisture availability period.

18.2 Agro-Climatic Zones of India by Planning Commission

The Planning Commission of India has divided the India into 15 broad agro-climatic zones on the basis of physiography, soils, geological formation, climate, cropping patterns and development of irrigation and mineral resources for extensive agricultural planning and developing future strategies. The main objective was to integrate plans of the agro-climatic regions with the state and national plans to enable policy development based on techno-agro-climatic considerations. The agro-climatic zone division will help to achieve the optimum utilization in a suitable manner within the framework of resource constraints and potentials of each region. This classification will also help in the development of efficient farming systems and crop improvement research programmes, and establishing principles for improved resource management. The agro-climatic zones in India are given below;

1. Western Himalayan Region

Area: Jammu and Kashmir, Himachal Pradesh and the hill region of Uttarakhand

Characteristics: The average temperature in July ranges between 5°C and 30°C, while in January it ranges between 5°C and -5°C. Mean annual rainfall varies between 75 cm to 150 cm. There is alluvial soil in the valleys of Kashmir, Kullu and

Dun, and brown soil in the hills. The valley floors grow rice, while the hilly tracts grow maize in the kharif season. Winter crops are barley, oats, and wheat. The region supports horticulture, especially apple orchards and other temperate fruits such as peaches, apricot, pears, cherry, almond, litchis, walnut, *etc.* Saffron is grown in this region. The main problems of this region are poor accessibility, soil erosion, landslides, inclement weather, inadequacy of marketing and storage facilities.

2. Eastern Himalayan Region

Area: Arunachal Pradesh, Assam, Sikkim, Meghalaya, Nagaland, Manipur, Mizoram, Tripura, and the Darjeeling district of West Bengal.

Characteristics: Temperature variation is between 25°C and 30°C in July and between 10°C and 20°C in January. Average rainfall is between 200-400 cm. The red-brown soil is not highly productive Jhuming (shifting cultivation) prevails in the hilly areas. The main crops are rice, maize, potato, tea. There are orchards of pineapple, litchi, oranges and lime. Infrastructural facilities in the region need to be improved and shifting cultivation controlled by developing terrace farming.

3. Lower Gangetic Plain Region

Area: West Bengal (except the hilly areas), eastern Bihar and the Brahmaputra valley

Characteristics: Temperature in July varies from 26°C to 41°C and for January from 9°C to 24°C. Average annual rainfall lies between 100-200 cm. Alluvial soils and are prone to floods. The region has adequate storage of ground water with high water table. Rice is the main crop which at times yields three successive crops (Aman, Aus and Boro) in a year. Jute, maize, potato, and pulses are other important crops. Planning strategies include improvement in rice farming, horticulture (banana, mango and citrus fruits), pisciculture, poultry, livestock, forage production and seed supply.

4. Middle Gangetic Plain Region

Area: Parts of Uttar Pradesh and Bihar.

Characteristics: The average temperature in July varies from 26°C to 41°C and that of January 9°C to 24°C average annual rainfall is between 100-200 cm. It is a fertile alluvial plain drained by the Ganga and its tributaries. Rice, maize, millets in kharif, wheat, gram, barley, peas, mustard and potato in rabi are important crops. Alternative farming systems, and utilising chaur lands for pisciculture are some measures to boost agricultural production. Reclamation of user lands, wastelands, and fallow lands for agriculture and allied activities (agro-forestry, silviculture, floriculture *etc.*) should be done.

5. Upper Gangetic Plains Region

Area: Central and western parts of Uttar Pradesh and the Hardwar and Udham Nagar districts of Uttarakhand.

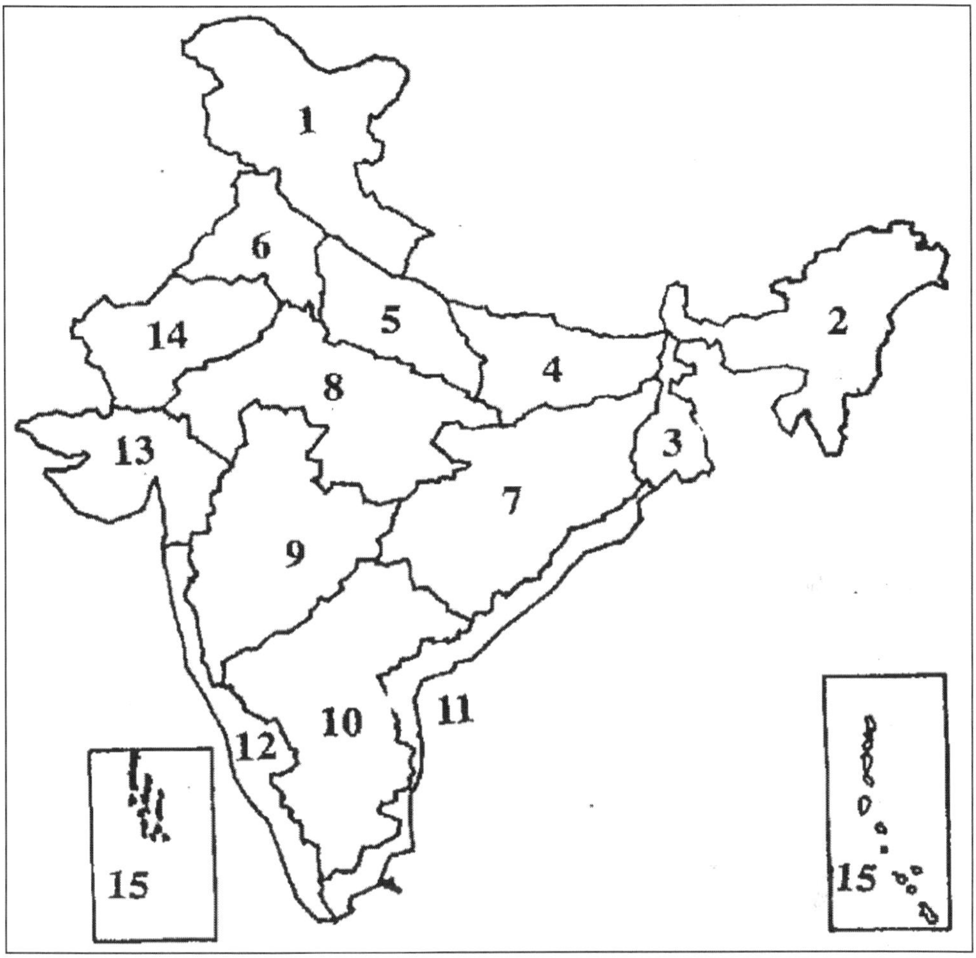

Figure 18.1. Agro-Climatic Zones of India.

1. Western Himalayan Region; 2. Eastern Himalayan Region; 3. Lower Gangetic Plain Region; 4. Middle Gangetic Plain Region; 5. Upper Gangetic Plains Region; 6. Trans Gangetic Plains Region; 7. Eastern Plateau and Hills; 8. Central Plateau and Hills; 9. Western Plateau and Hills; 10. Southern Plateau and Hills; 11. Eastern Coastal Plains and Hills; 12. Western Coastal Plains and Ghats; 13. Gujarat Plains and Hills; 14. Western Dry Region; 15. Island Regions.

Characteristics: The climate is sub-humid continental with temperature in July between 26°C to 41°C and temperature in January between 7°C to 23°C. Average annual rainfall is between 75-150 cm. The soil is sandy loam. Canal, tube-well and wells are the main source of irrigation. This is an intensive agricultural region wherein wheat, rice, sugarcane, millets, maize, gram, barley, oilseeds, pulses and cotton are the main crops. Besides modernising traditional agriculture, the region needs special focus on dairy development, horticulture and developing multiple mixed cropping patterns.

6. Trans Gangetic Plains Region

Area: Punjab, Haryana, Chandigarh, Delhi and the Ganganagar district of Rajasthan.

Characteristics: Semi-arid characteristics prevail over the region, with July's mean monthly temperature between 25°C and 40°C and that of January between 10°C and 20°C. The average annual rainfall varies between 65-125 cm. The soil is alluvial which is highly productive. Canals and tube-wells and pumping sets have been installed by the cultivators and the governments. The intensity of agriculture is the highest in the country. Important crops include wheat, sugarcane, cotton, rice, gram, maize, millets, pulses and oilseeds *etc.* The region has the credit of introducing green revolution in the country and has adopted modern methods of farming and mechanization. The region is facing the menace of water logging, salinity, alkalinity, soil erosion and falling water table. This area are highest net sown area, irrigated area and cropping intensity.

7. Eastern Plateau and Hills Region

Area: Chhotanagpur Plateau, extending over Jharkhand, Orissa, Chhattisgarh.

Characteristics: The mean temperature 26°C to 34°C in July, 1°C to 27°C in January. The average annual rainfall varies between 80-150 cm. The topography is undulating with a slope of 1-10 per cent. Soils are red and yellow with occasional patches of laterites and alluviums. The region is deficient in water resources due to plateau structure and non-perennial streams. Rainfed agriculture is practiced growing crops like rice, millets, maize, oilseeds, ragi, gram and potato. Steps to improve agricultural productivity and income include cultivation of high value crops of pulses like tur, groundnut and soyabean. On upland rain-fed areas, growing crops like urad, castor, and groundnut in kharif and mustard and vegetables in irrigated areas, improvement of indigenous breeds of cattle and buffaloes, extension of fruit plantations, renovation including desilting of existing tanks and excavation of new tanks, development of inland fisheries in permanent water bodies and adopting integrated watershed development approach to conserve soil and rain water.

8. Central Plateau and Hills Region

Area: Madhya Pradesh, Uttar Pradesh, Rajasthan.

Characteristics: Semi-arid climatic conditions prevail over the region with temperature in July 26°C to 40°C, in January 7°C to 24°C. The average annual rainfall from 50-100 cm. Soils is mixed red, yellow and black. There is scarcity of water. Crops grown are millets, wheat, gram, oilseeds, cotton and sunflower. In order to improve agricultural returns, measures to be adopted are water conservation through water saving devices like sprinklers and drip system; dairy development, crop diversification, ground water development, reclamation of ravine lands. About 75 per cent area comes under rainfed. Land topography is highly variable and about 30 per cent land is not available for cultivation.

9. Western Plateau and Hills Region

Area: Deccan plateau (Maharashtra), Madhya Pradesh and Rajasthan.

Characteristics: It is a region of the regur (black) soil with July temperature between 24°C and 41°C, January temperature between 6°C and 23°C. The average annual rainfall of 25-75 cm. Wheat, gram, millets, cotton, pulses, groundnut and oilseeds are the main crops in the rain-fed areas, while in the irrigated areas, sugarcane, rice and wheat are cultivated. Also grown are oranges, grapes and bananas. Attention should be paid to increasing water efficiency by popularizing water saving devices like sprinklers and drip system. The lower value crops of jowar, bajra and rainfed wheat should give way to high value oilseeds. Five per cent area under rain-fed cotton and jowar could be substituted with fruits like ber, pomegranate, mango and guava. Improvement of milk production of cattle and buffalo through cross-breeding along with poultry development should be encouraged.

10. Southern Plateau and Hills Region

Area: Interior Deccan and includes parts of southern Maharashtra, the greater parts of Karnataka, Andhra Pradesh, and Tamil Nadu uplands from Adilabad district in the north to Madurai district in the south.

Characteristics: The mean monthly temperature of July varies between 25°C and 40°C, and the mean January temperature is between 10°C and 20°C. Annual rainfall is between 50-100 cm. It is an area of dry-zone agriculture where millets, oilseeds, and pulses are grown. Coffee, tea, cardamom and spices are grown along the hilly slopes of Karnataka plateau. Some of the area under coarse cereals may be diverted to pulses and oilseeds. Horticulture, dairy development and poultry farming should be encouraged.

11. Eastern Coastal Plains and Hills Region

Area: Coromandal and northern coasts of Andhra Pradesh, Orissa and Tamil Nadu.

Characteristics: The mean July temperature ranges between 25°C and 35°C and the mean January temperature varies between 20°C and 30°C. The mean annual rainfall varies between 75-150 cm. The soils are alluvial, loam and clay and are troubled by the problem of alkalinity. Main crops include rice, jute, tobacco, sugarcane, maize, millets, groundnut and oilseeds. Main agricultural strategies include improvement in the cultivation of spices (pepper and cardamom) and development of fisheries. These involve increasing cropping intensity using water-efficient crops on residual moisture, discouraging growing of rice on marginal lands and bringing such lands under alternate crops like oilseeds and pulses; diversifying cropping and avoiding mono-cropping; developing horticulture in upland areas, social forestry and dairy-farming. About 70 per cent land area comes under rainfed.

12. Western Coastal Plains and Ghats Region

Area: Malabar and Konkan coastal plains and the Sahyadris, (Maharashtra, Tamil Nadu, Kerala, Karnataka, Goa).

Characteristics: The region is humid with the mean July temperature varying between 25 to 30°C and mean January temperatures between 18 to 30°C. The mean annual rainfall is more than 200 cm. The soils are laterite and coastal alluvial. Rice, coconut, oilseeds, sugarcane, millets, pulses and cotton are the main crops. The region is also famous for plantation crops and spices which are raised along the hill slopes of the Western Ghats. The agricultural development must focus attention on raising of high value crops (pulses, spices and coconut). Development of infra-structural facilities and promotion to prawn culture in brackish water should be encouraged.

13. Gujarat Plains and Hills Region

Area: Hills and plains of Kathiawar and the fertile valleys of Mahi and Sabarmati rivers.

Characteristics: It is an arid and semi-arid region with the mean July temperature reading 30°C and that of January about 25°C. The mean annual rainfall varies between 50-100 cm. Soils are regur in the plateau region, alluvium in the coastal plains, and red and yellow soils in Jamnagar area. Groundnut, cotton, rice, millets, oilseeds, wheat and tobacco are the main crops. It is an important oilseed producing region. The main strategy of development in this region should be canal and groundwater management, rain water harvesting and management, dry land farming, agro-forestry, waste land development and developing marine fishing and brackish water aquaculture development in coastal zones and river deltas.

14. Western Dry Region

Area: West of Aravalli (Rajasthan).

Characteristics: The desert climate further causes high evaporation and contrasting temperatures 28°C to 45°C in June and 5°C to 22°C in January. Bajra, jowar, and moth are main crops of kharif and wheat and gram in rabi. Livestock contributes greatly in desert ecology. The main areas needing a thrust for development are rainwater harvesting, increasing yield level of horticultural crops like water melon, guava and date palm, adopting high quality germplasm in cattle to improve their breed; and adopting silvipastoral system over wastelands.

15. Island Region

Area: Andaman-Nicobar and Lakshadweep.

Characteristics: The region have typically equatorial climate. The annual rainfall is 300 cm spend over 8-9 month. The mean July and January temperature of Port Blair being 30°C and 25°C respectively. The soils vary from sandy along the coast to clayey loam in valleys and lower slopes. The main crops are rice, maize, millets, pulses, arecanut, turmeric and cassava. Nearly half of the cropped area is under coconut. The area is covered with thick forests and agriculture is in backward stage. The main thrust in development should be on crop improvement, water management and fisheries. Improved variety of rice seeds should be popularized so as to enable farmers to take two crops of rice in place of one. For fisheries development multi-purpose fishing vessels for deep sea fishing should be introduced, suitable infrastructure for storage and processing of fish should be built up, and brackish water prawn culture should be promoted in the coastal areas.

18.3 Agro-Climatic Zones Under NARP

The National Agricultural Research Project (NARP) was launched by ICAR for initiating agricultural research in the agro-climatic zones of the country. The objective was to set up or upgrade a zonal research station in each agro-climatic zone for generating location specific, need based research targeted for specific agro-ecological situations. The focus was on analyzing agro-ecological conditions and cropping patterns and come out with a programme directly targeted to solve the major bottle necks of agricultural growth in a zone based on natural resources, major crops, farming systems, production constraints and socio-economic conditions prevalent in that zone. Stress was on technology generation. In NARP, the country was divided into 127 agro-climatic zones.

18.4 Agro-Ecological Regions of India

Agro-ecology refers to the study of agricultural ecosystems and their components as they function in themselves and as a part of the larger ecosystem. Agro-ecological zone is the land unit carved out of agro-climatic zone and superimposed on landform which influences both the climate and length of growing period. The National Bureau of Soil Survey and Land Use Planning, Nagpur (NBSS and LUP) came up with twenty agro-ecological zones based on the growing period as an integrated criteria of effective rainfall, soil groups, delineated boundaries adjusted to district boundaries with a minimal number of regions.

Arid Ecosystem

1. Western Himalayas, cold eco-region with shallow and skeleton soils, LGP <90 days. Station: Ladakh.
2. Western plain Kachohh and parts of Kathiawar Peninsula, hot arid eco-region, desert and saline soils. LPG <90 days. Station: Ladakh.
3. Deccan plateau, hot arid ecoregion, red and black soils. LGP <90 days. Semiarid ecosystem. Station: Bellary, Karnataka.
4. Northern plain and central high lands, hot semiarid eco-region, alluvial soils. LGP 90-150 days. Station: Aligarh, UP.
5. Central high lands, Gujarat plains, Kathiawar peninsula, hot semiarid eco-region, medium and deep black soils. LGP 90-150 days. Station: Vadodara, GJ.

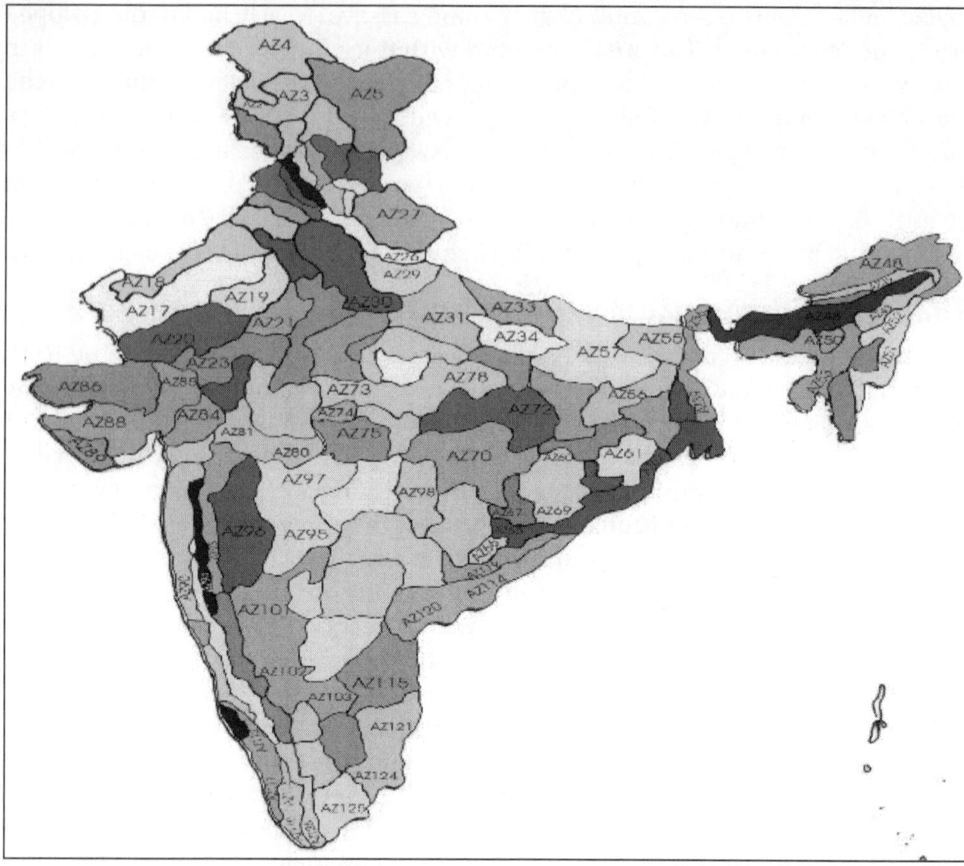

Figure 18.2: Agro-climatic Zones under NAR.

6. Deccan plateau, hot semiarid ecoregion. LGP 90-150 days. Station: Akola.
7. Telangana, Eastern Ghats, hot semiarid eco-region. LGP 90- 150 days. Station: Hyderabad.
8. Eastern Ghats, Tamil Nadu uplands and Karnataka plateau, hot semiarid eco-region. LGP 90-150 days. Station: Tiruchirapalli, Kerala.

Sub-humid Ecosystem

9. Northern plain, hot sub-humid (dry) eco-region, red and black soils. LGP 150-180 days. Station: Patna.
10. Central highlands, hot sub-humid eco-region, black and red soils. LGP 150-180 (210) days.
11. Eastern plateau, hot sub-humid eco-region, red and yellow soils, (210) days. LGP 150-180 days.
12. Eastern plateau (Chotanagpur) and Eastern Ghats hot sub-humid eco-region, red and lateritic soils. LGP 150-180 (210) days.

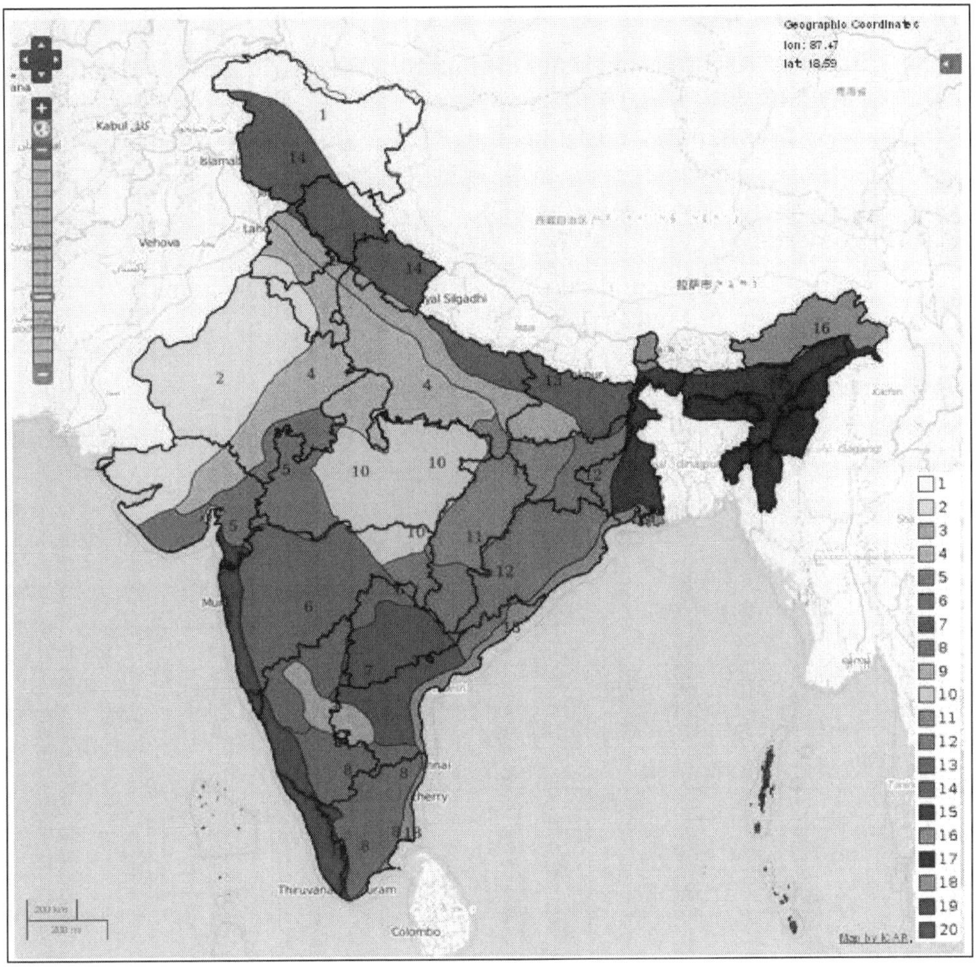

Figure 18.3. Agro-Ecological Regions of India.

13. Eastern plain, hot sub-humid (moist) eco-region, alluvial soils. LGP 180-210 days.
14. Western Himalayas, warm sub-humid to humid eco-region with brown forest soils. LGP 180-210+ days.

Humid-Perhumid Ecosystem

15. Bengal and Assam plain hot sub-humid (moist) to humid eco-region, alluvial soils. LGP 210+ days.
16. Eastern Himalayas, warm per-humid eco-region, brown and red hill soils. LGP 210 + days.
17. North eastern hills, warm per-humid eco-region, red and lateritic soils. LGP 210+ days.

Coastal Ecosystem

18. Eastern coastal plain, hot sub-humid to semiarid eco-region, coastal alluvium. LGP 90-210 +days.
19. Western Ghats and coastal plain, hot humid-per-humid eco-region, red, lateritic and alluvium derived soils. LGP 210+ days.

Island Ecosystem

20. Andaman Nicobar and Lakshadweep, hot humid to per-humid eco-region, real loamy and sandy soils. LGP 210+ days.

18.5 Agro-Climatic Zones of Maharashtra

The Maharashtra state has been divided into nine agro-climatic zones based on rainfall, soil type and vegetation are mentioned belows,

Sl.No.	*Agro-climatic zone*	*Head Office*
1	South Konkan Coastal Zone	Vengurla
2	North Konkan Coastal Zone	Karjat
3	Western Ghat Zone	Igatpuri
4	Sub Mountain Zone	Kolhapur
5	Western Maharashtra Plain Zone	Pune
6	Western Maharashtra Scarcity Zone	Solapur
7	Central Maharashtra Plateau Zone	Aurangabad
8	Central Vidarbha Zone	Yavatmal
9	Eastern Vidarbha Zone	Sindewadi

1. South Konkan Coastal Zone

It comprises districts of Ratnagiri, south Raigad and Sindhudurg. The zone has an undulating topography with bills and rocky plains alternating.

Climate: The zone is characterized by warm and humid climate. The bulk of the rainfall is received during June to September from southwest monsoon. The average annual rainfall is 3100 mm in about 101 rainy days.

Soils: Lateritic soil is the predominant type of soil in this zone. While along the sea coast, saline and alluvial soils are occurred. The intense leaching due to high rainfall has removed the bases from the soil profile giving it a distinctly acidic reaction.

Crops and cropping patterns: Paddy is the major cereal crop and Ragi is the second important crop of the zone. Vari a minor millet is grown on hill slopes on low fertile soil during *kharif*. Oil seed crops like niger, sesamum are also grown in *kharif* on limited areas. Due to typical soil and climate, the zone is ideally suited for horticultural crops *viz*, mango, coconut, arecanut, cashewnut, jackfruit, banana, pineapple *etc.*

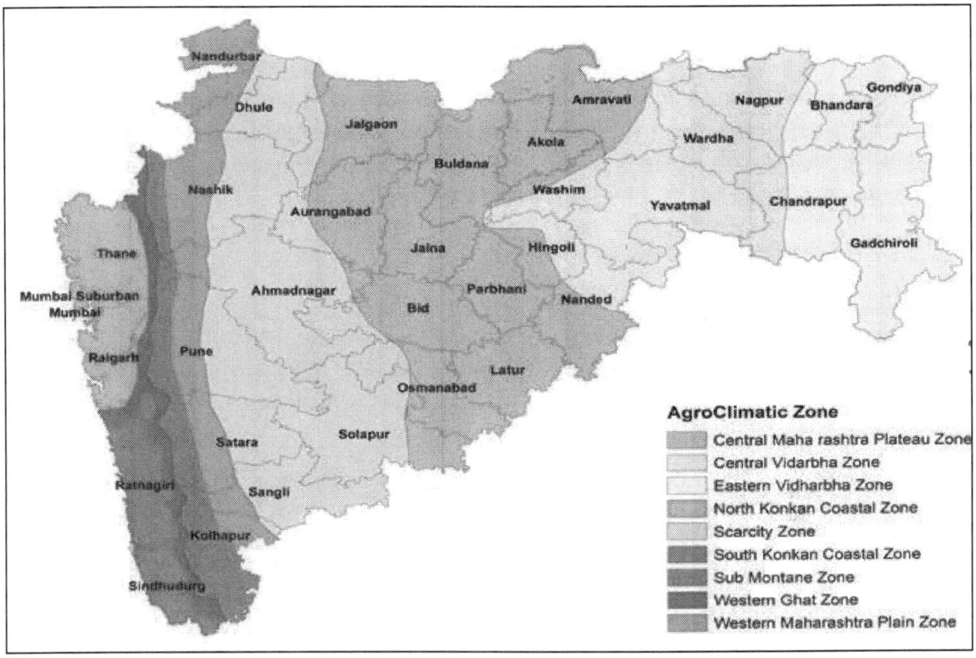

Figure 18.4. Agro-Climatic Zones of India, Maharashtra.

2. North Konkan Coastal Zone

This zone comprising districts of Thane and Raigad.

Climate: The zone is characterized by warm and humid climate. The average rainfall is 2600 mm in 87 rainy days.

Soils: The predominant soils found in the Thane and northern part of Raigad districts are non-lateritic coarse and shallow. The soils are neutral to acidic in reaction medium black soils occur in patches in the northern part of the zone. Southern part of Raigad has lateritic soils.

Crops and cropping patterns: Paddy is the major crop in the area. Ragi and Vari are also important crops. Major fruit crops are banana and sapota in Thane and mango in Raigad.

3. Western Ghat Zone

It includes hilly and high lying terrains around the Ghats of Amboli, Phonda and Amba in Kolhapur, Koyna, Mahabaleshwar in Satara, Lonawala, Khandala in Pune, Kalsoobai hills in Akola of Ahmednagar district and Igatpuri, Trimbakeshwar in Nasik district.

Climate: Rainfall at high lying terrains ranges between 3000 to 6000 min and the area is mainly covered by deciduous to semi deciduous forests.

Soils: The soils are mostly light lateritic and reddish brown soil distinctly acidic and poor in fertility.

Crops and cropping patterns: The principal crops grown are paddy, ragi, vari, *kharif* sorghum, groundnut, niger, gram *etc.* The sugarcane is grown as cash mop. The horticulture is the priority need of this zone. The area is suited for fruits like mango, cashewnut, jackfruit, kavanda, jamun.

4. Sub-Mountain Zone

This zone is located on the eastern slopes of Sahyadri ranges. It includes surgana, peth and Nasik talukas of Nasik district, Rajgurunagar, Maya, Velha and Bhor talukas of Pune, Javali, Patan and Karad talukas of Satara, Shahuwadi, Panhala, Karvir, Kagal, Bhudargad, Ajara, Gadhinglaj talukas of Kolhapur and Shirala taluka of Sangali.

Climate: The annual rainfall of the zone ranges between 1700-2500 mm which is received mostly through southwest monsoon. The climate is characterized by relatively hot summer, humid rainy season and moderately cold winters.

Soils: The soils are reddish brown to black tending towards lateritic nature with varying depths and textures.

Crop and cropping patterns: The cropping pattern is mainly dominated by *kharif* cereals like paddy, sorghum, pearl millet, maize, ragi followed by groundnut and sugarcane. During Rabi, sorghum, wheat, gram are the important crops. Potato, chillies, tomato, brinjal are important vegetable crops being grown in the area, Guava, mango, banana and cashew nut are the important fruit crops.

5. Western Maharashtra Plain Zone

This zone includes areas in western tehsils of Ahmednagar and Sangli and Central tahsils of Nasik, Pune, Satara and Kolhapur.

Climate: This zone has well distributed rainfall ranging from 950-1250 mm. The maximum rainfall is received from southwest monsoon from June to October. The period of water availability ranges from 120-150 days.

Soils: The topography is plain in the zone. The soils are predominantly greyish black with varying texture and depth. The soils are medium deep, dark brown in colour, clay loam in texture and moderately alkaline.

Crop and cropping patterns: The principal crops are *kharif* and *rabi* sorghum, groundnut, soybean, pearl millet, wheat, sugarcane, green gram, black gram, red gram, ragi, chickpea under irrigated condition.

6. Western Maharashtra Scarcity Zone

This region includes Solapur, Ahmednagar parts of Satara, Sangli, Pune, Kolhapur, Dhule, Jalgaon, Nasik, Aurangabad, Jalna, Bhid and Osmanabad districts.

Climate: The annual rainfall is less than 750 mm which receives in 45 rainy days. Two peaks in rainfall are observed, first during June and July and second during September, resulting in bimodal pattern of rainfall distribution. The dry spells are experienced during July and August. The water availability periods vary from 60 to 140 days and the PE value is more than 1800 mm.

Soils: The soils are vertisol with base saturated. They have montmorilonite type clay, which swells when wet and shrink on drying producing deep cracks. They are low in nitrogen, low to medium in phosphate but rich in potash. Shallow soils are suited for *kharif* cropping and deep soils for *rabi* cropping.

Crops and Cropping patterns: *Kharif* crops like pearl millet, sorghum, pulses, groundnut, sunflower are taken on shallow soils. In heavy soil sorghum, safflower, chickpea, sunflower are taken in *rabi* season.

7. Central Maharashtra Plateau Zone

This zone comprises of parts of Aurangabad, Jalna, Bhid and Osmanabad districts, major parts of Parbhani and Nanded districts, Latur district. This zone includes entire Buldhana and Akola districts excluding southern parts of Washim and Mangrul pir talukas, whole of Amravati, excluding eastern part of Chandrapur and Morshi talukas, Darwha and Pusad talukas of Yevatmal district, whole Eastern part of Jalgaon district, Shirpur taluka of Dhule, South Solapur and akkalkot talukas of Solapur.

Climate: The annual rainfall is 900 to 950 nun with 35 to 65 rainy days. In monsoon (June-September), more than 75 per cent of rainfall is received in all the district of this assured rainfall zone.

Soils: Soils are derived from Deccan trap and range from black to red colours. The soils are vertisols, entisols and inceptisol.

Crops and cropping patterns: The sorghum is the predominant crop of the region also cotton and soybean are cultivated. The oilseeds (groundnut, sunflower) and pulses (green gram, black gram, red gram, gram).

8. Central Vidharbha Zone

This zone includes entire Wardha district, major part of Nagpur and Yevatmal districts, part of Western Chandrapur district, parts of Anrangabad, Parbhani, Jalna and Nanded.

Climate: The average annual rainfall ranges between 950 to 1250 mm which received in about 59 rainy days.

Soils: Soils of the zone are derived from basalt rocks, black in colour and have varying depths. These soils are medium to heavy in texture, fairly high in clay content, alkaline in reaction, high time reserves with high base saturation of the exchange complex. Vertisols swells after wetting and shrinks after drying producing deep and wide cracks.

Crops and cropping patterns: The cropping patterns involves cotton, *kharif* sorghum, Red gram, wheat, pulses and oilseeds, paddy and rabi sorghum are also included in the rotation.

9. Eastern Vidharbha Zone

This zone includes entire Bhandara and Gadchiroli districts, eastern part of Chandrapur district and Umred tahsil of Nagpur district.

Climate: This zone receives average annual rainfall of 1250 to 1700 mm in 60-70 days.

Soils: Soils are derived from mixed parent rocks like granite, geneses and schists. They are yellowish brown to red in colour and areas in the south and west have medium black soils.

Crops and cropping patterns: Paddy is the predominant crop the zone in Bhandara district during *kharif*, followed by pulses gram in rabi. In Bramhpuri, and Sironcha taluks of Chandrapur district, paddy although is predominant crop is associated with *rabi* sorghum and pulse and oilseeds during *rabi* season. In Umred taluka of Nagpur district almost six crops including rabi sorghum, wheat, *kharif* sorghum, chillies, paddy and linseed together form the major crop pattern with greater area under rabi crops.

Chapter 19

Agrometeorological Observatory

19.1 Introduction

An observatory is a specially designed station or place where the regular and simultaneous records of the weather data are made by physical measurements using various techniques, sensors, skills, recorders, instruments *etc.* by standard methods at hours recommended by IMD and WMO. Meteorological observatory is a place where all the necessary instruments are exposed for measuring weather phenomenon.

19.2 Classification of Agro-met Observatories

World Meteorological Organization has divided Agro meteorological observatories into four categories;

a) Principal Agricultural Meteorological Stations

These provide detailed simultaneous meteorological and biological information and it is where research in agricultural meteorology is carried out. The instrumental facilities, range and frequency of observations, in both meteorological and biological are such that fundamental investigations into agricultural meteorological questions of interest to the countries or regions concerned can be carried out.

b) Ordinary Agricultural Meteorological Stations

These provide on a routine basis, simultaneous meteorological and biological information. These are also equipped to assist in research into specific problems. The programme of biological or phenological observations for research will be related to the local climatic regime of the station and to local agriculture.

c) Auxiliary Agricultural Meteorological Stations

These also provide meteorological and biological information. However, to a lesser extent than the above two. The meteorological information includes such items as soil temperature, soil moisture, potential evapotranspiration, duration of vegetative wetting, and detailed measurements in the very lowest layer of the atmosphere. The biological information may cover phenology, onset and spread of plant diseases, and so forth.

d) Agricultural Meteorological Station for Specific Purposes

These are the stations set up temporarily or permanently for the observation of one or several variables and/or specified phenomena.

19.3 Point Should be Considered While Selecting Site of Agro-met Observatory

- The location should be at the center of the farm in a bare plot of 55m long (North-South) and 36m wide (East-West).
- The site should be perfectly leveled.
- The site should be free from water logging.
- It should be well exposed to atmosphere.
- It should be easily approachable by roads so as to avoid difficulty in recording observations.
- It should be away from the hills forest irrigation channels drains tank and water reservoir.
- It should be representative of soil type of that area.
- There should not be a large area of concrete building or tree near the plots.
- The distance should be more than 2-3 times the height of the tallest object.
- All sides, Eastern and western side of the site should be free from big structure. The distance should be more than 10 times the height of tallest structure if any.
- Suitable locking arrangements should be provided for safety of the instrument.

19.4 Plan of Layout

The whole area of the observatory should be fenced with the help of iron poles and barbed wire. The distance between two iron poles should be 3 meters. The height of iron poles should be 1.25 m above ground level and 0.75 m should be buried in the soil. The barbed wire should be stretched with 5 strands and one cross strand. Thus, for east-west 13 and North-south 17 poles will be required, excluding 2 supporting poles for gate. A gate of 1.25 m x 0.75 m dimensions should be fixed in the fencing at appropriate place depending on suitability of approach.

The plot should be divided in three equal division breadth wise of 12 m each. On second- and third-line various instruments are installed by keeping 9 m distance between two instruments. The sunshine recorder and pan evaporimeter are installed on the central line 18 m in front of each other about 5 m away from the fencing at the two ends. Instruments are installed in a descending order of heights from North to South. The instruments standardized by IMD should only be used. Agro-meteorological observatory has the dimension of 55 m x 36 m. It is divided by Line AB, Line CD and Line EF (N-S).

Line AB

It consists of soil thermometer leaving 14 m distance from southern boundary of the observatory. Wind vane is installed at a distance of 14m from northern boundary. Ordinary rain gauge and single Stevenson screen are installed by keeping 9 m distance from soil thermometer and wind vane, respectively.

Line EF

It is a central line of observatory. It consists of evaporimeter installed at 5 m distance from South boundary. From North boundary at 5 m distance there is a sunshine recorder and at the center of line microclimate pole is installed. Grass minimum thermometer is installed by keeping 14 m distance from South boundary. This line is 18 m away from east boundary.

Line CD

At a distance of 14 m from North boundary there is cup counter anemometer. The double Stevenson screen and self-recording rain gauge are placed at a distance of 9 m from anemometer and dew gauge stand, respectively. This line is 12 m away from east boundary. Dew gauge is installed by keeping 14m distance from South boundary.

19.5 Standard of Time

Greenwich Mean Time (GMT)

It is the mean solar time at the Royal Observatory in Greenwich, London reckoned from midnight (00.00° Longitude). The standard of time adopted for reporting special of these observation is always the Greenwich Mean Time which is 05:30 hours less than the corresponding India Standard Time. It is known as Universal Time.

Indian Standard Time (IST)

The IST is controlled from Allahabad with longitude 82.5°E which is taken as standard longitude for India.

Local Mean Time

This is the local time based on the transit of the Sun. The LMT corresponding IST changes from place to place depending upon the longitude of a place. When

Sun rises at any place, it is said to 7 o clock locally. At Allahabad observation are recorded at 0700 hrs IST which is LMT at that place. For calculation of LMT, for every place East of Allahabad for one-degree differences of longitude (111 km) 4 minutes are deducted from IST to get LMT of that place, while for a place one degree longitude west to Allahabad 4 minute are added in IST to get LMT as sunrise will be 4 minutes later for that place.

19.6 Installation of Meteorological Instruments

Sl.No.	*Instrument Position*	*Height*	*Remarks*
1	U.S.W.B Open Pan Evaporimeter	0.30 m	☆ Its base should be placed on a wooden platform ☆ Wooden platform should be perfectly horizontal. ☆ Under the wooden platform the ground should be free from weeds for the movement of the air. ☆ Diameter of pan is 120.7 cm.
2	Thermo-hygrograph	1.2 m	☆ It should be placed in a double layered wooden screen known as Stevenson's screen. ☆ It should face towards north in the northern hemisphere.
3	Microclimate Post	3.6 m	☆ It should be at 9-10m distance from the sunshine recorder.
4	Sunshine recorder	3.05 m	☆ It should be installed on a concrete pillar with steps. ☆ Base should be perfectly horizontal ☆ The groove of the recorder should be oriented in E-W direction. ☆ Diameter of glass globe is 10 cm.
5	Soil thermometers		☆ Support the thermometers with iron stand and protect them with rod fence ☆ The bulb of the thermometers should be placed at a desired depth, keeping the stem at 60° angles from the horizontal

Sl.No.	*Instrument Position*	*Height*	*Remarks*
6	Dew gauge Post		✰ It should be 9-10m away from soil thermometers towards south ✰ The post should be vertical with horizontal arms at different heights.
7	Self-Recording Raingauge	0.75 m	✰ It should be fixed on a concrete platform ✰ The rim of the funnel should be horizontal ✰ It should be 2-3 m away from an ordinary rain gauge.
8	Ordinary Rain gauge	0.30 m	✰ It should be fixed on a concrete platform ✰ The rim of the funnel should be horizontal ✰ It should be between Stevenson screen and the self-recording rain gauge.
9	Stevenson Screen	1.22 m	✰ Thermometers should be kept within Stevenson screen ✰ Maximum and minimum thermometers should be placed horizontally with their bulbs tilted slightly downwards ✰ Dry and wet bulb thermometers should be placed vertically ✰ Bulb of the wet bulb thermometer should be covered with a muslin cloth and it should be kept wet by a thread, one end of which is fixed to the muslin and other end is kept in the distilled water in a bottle.
10	Wind vane	3.05 m	✰ It is installed on a platform fixed at the top of a pillar which is at a distance of 9-10m from screen.
11	Cup Anemometer	3.05 m	✰ It is installed on a platform fixed at the top of a pillar to the south of the Stevenson's screen.

19.6 Instruments used in Observatory, and Time and Unit of Observation

Weather Parameter	*Instrument Used*	*Unit of Measurement*	*Time of Observation*
Sunshine Hours	Campbell Stokes Sunshine Recorder	Hours	Before sunrise card is to be inserted and after sunset cad is to be removed
Wind Velocity	Robinsons Cup Anemometer, Pilot Balloon, Areovane	km/hr, miles/ hr, meter/sec, knot	0830 hours1730 hours
Wind Direction	Aerovane, Windvane	Degree, all direction	0830 hours
Relative Humidity	Thermometrs, Assmanns Psychrometer, Hair Hygrograph	Percentage	0800 hours0900 hours
Soil Temperature	Soil Thermometer	°C,°F,°K	0700, 0830, 1400,1730 hours
Air Temperature	Sling Thermometer, Thermograph, Thermometr	°C,°F,°K	0830 hours1730 hours
Evaporation	Floating Pan Evaporimeter,Open Pan Evaporimeter	mm, cm. inches	0830 hours1730 hours
Rainfall	Ordinary Type Rain Gauge,Automatic Type Rain Gauge	mm, cm, inches	0830 hours1730 hours
Atmospheric Pressure	Simple Barometer, Fortin's Barometer, Aneroid Barometer, Barograph	Milibar, Atmosphere, kpa, inch	0830 hours, 1130 hours, 1430 hours, 1730 hours
Cloud Cover	Visval Observations	Okta (0-8)	-

Chapter 20

Important Question

- ☆ Heat flow in air by process of radiation.
- ☆ Heat flow in solid/soil takes place mainly through the process of Conduction.
- ☆ The transfer of heat by molecules-to-molecules contact is known as conduction.
- ☆ Heat flow in liquid/water by process of convection.
- ☆ The mean sea level pressure is 1013.25 millibars.
- ☆ 0.4-0.7 nm is the wavelength of visible solar radiation.
- ☆ Wavelength longer than 750 m/μ is not visible to the eye are called infrared.
- ☆ Blue colour of the sky and red colour of sunset is due to dispersion.
- ☆ Blue color of the sky is due to rayleigh scattering.
- ☆ Radio transmission found in Ionosphere zone.
- ☆ The value of solar constant is 1.94 $cal/cm^2/min$.
- ☆ De Condole (1900) is the first scientist attempted to classify the climate.
- ☆ Koppen and Thornthwaite classified the climate on the basis of annual rainfall and annual evaporation.
- ☆ Troll classified the climate on the basis of humid month and temperature.
- ☆ The percentage concentration of CO_2 in air and soil is 0.03 per cent and 0.30 per cent.
- ☆ Auguste Mouchout discovered solar energy.
- ☆ The temperature on the sun is around 6000°C.

- ☆ Monsoon is an Arabic word.
- ☆ The response of plants to the relative length of day and night is known as photoperiodism.
- ☆ The first meteorological satellite was launched in 1960 year.
- ☆ The rate of the rotation of the earth determines the strength of the Coriolis force.
- ☆ The cup anemometer rotates in west to east direction.
- ☆ According to planning Commission, the country has been divided in 15 agro-climatic regions.
- ☆ Movement of air occurs in the atmosphere upto the height of 10 km.
- ☆ The normal lapse rate is 6.5°C or 3.5°F/1000ft.
- ☆ Condensation is the process by which a cloud droplets first form.
- ☆ Clouds are classified and named according to their latitude and form or appearance.
- ☆ A cumulus clouds is recognized mainly by its vertical heights.
- ☆ Tornado most often move towards North to East direction.
- ☆ Cumulonimbus clouds is most common in a hurricane.
- ☆ The vertical motions that occur when the air is unstable is termed as convection.
- ☆ The term latent means hidden.
- ☆ An increase in albedo would be accompanied by decrease in radiative equilibrium temperature.
- ☆ The charge particles from the sun that travel through space at high speed are called solar wind.
- ☆ Relative humidity is the percentage of water vapour present in the air compared to that required for saturation.
- ☆ A dim, watery sun visible through a grey sheet like cloud layer is often good indication of Altostratus cloud.
- ☆ The combined albedo of the earth and the atmosphere is approximately 30-34 per cent.
- ☆ Quantum sensor is used to measure PAR.
- ☆ Sensor used in quantum sensor is photodiode.
- ☆ Sensor used in albedometer is thermopile.
- ☆ The line joining the places having equal temperature is called as isotherm.
- ☆ The line joining the places having equal atmospheric pressure is called as isobar.
- ☆ The line joining the places having equal amount of rainfall is called as isohyet.

- ☆ The line joining the places having equal duration of sunshine hours is isohel.
- ☆ USWB stands for United states weather bureau.
- ☆ Local time is also called as solar time.
- ☆ NCMRWF stands for national center for medium range weather forecasting.
- ☆ IMD stands for India meteorological department and started in 1875.
- ☆ MWO stands for world meteorological organization and started in 1951.
- ☆ Indian standard time is calculated from the longitude of Allahabad (82.50°E).
- ☆ Ozonosphere is called as umbrella of the earth.
- ☆ The ozone layer was discovered in 1913 by Charles fabry and henri buisson.
- ☆ The term acid rain was coined in 1872 by Robert Augus Smith.
- ☆ Latitude is an angular distance of a place from equator of the earth.
- ☆ Altitude is the height of a place from mean sea level.
- ☆ Temperature gradient is the rate of change of temperature with a horizontal distance.
- ☆ Contour is the lines joining point of equal elevations.
- ☆ Isopleths is the line joining uniform values of any phenomena.
- ☆ Isohyts the lines joining point of equal amounts of rainfall.
- ☆ Isobaths is the lines joining point of equal water table elevations.
- ☆ Isopluvial is the lines joining point of equal depth of rainfall.
- ☆ Isohels is the lines joining point of equal sunshine hours.
- ☆ Isophene is lines joining point of equal similar seasonal phenomenon.
- ☆ Isotherm is lines joining point of equal temperature.
- ☆ Isotech is lines joining point of equal wind velocities.
- ☆ The belt of maximum insolation is called thermal equator.
- ☆ Any biochemical reaction increases with increase in temperature, for every 10°C increase in temperature the rate of reaction may double of even be more it is called Quotient 10 or Q10.
- ☆ At 24°C temperature is net carbon dioxide uptake in barley and wheat.
- ☆ Phenology is a science which deals with the reoccurrence of important phases of animals and vegetables life in relation to climate during the year.
- ☆ Climograph is a chart shows a places yearly climate pattern.
- ☆ Climate water budgeting is given by Thornthwaite and Matter.

- ☆ Zenith is imaginary point directly above particular locations.
- ☆ Storm surge is the force of cyclone decreases with increases in diameter from the sea. so, coastal areas often struck by severe cyclonic storms, this resulted in abnormal rise in the sea level.
- ☆ Surge is generated due to interaction of air, sea and land.
- ☆ Seismology is the scientific study of earthquakes.
- ☆ The total amount of heat required, between the lower and upper thresholds, for an organism to develop from one point to another in its life cycle is calculated in units called degree-days or heat units.
- ☆ NCMRWF stands for National Centre for Medium Range Weather Forecasting.
- ☆ IPCC stands for Intergovernmental Panel on Climate Change.
- ☆ ICRISAT stands for International Crops Research Institute for the Semi-Arid Tropics.
- ☆ IITM stands for Indian Institute of Tropical Metrology.
- ☆ Carbon dioxide constitutes about 0.03 per cent of dry air and efficient absorber long wave radiation from the earth.
- ☆ Aerosols is responsible for regional cooling effect.
- ☆ Chlorofluorocarbons is responsible for ozone depletion.
- ☆ Ozone is responsible for depletion of solar radiation.
- ☆ The atmosphere is transparent to the spectral region between 8.5 to 11u is known as atmospheric window.
- ☆ The climate normal's are generally worked out for a period of 30 years.
- ☆ The IMD prepares climate normal's for every 30 years period since 1901.
- ☆ NWP stands for numerical weather prediction.
- ☆ Dr. L. A. Ramdas is known as the father of Agricultural Meteorology in India.
- ☆ The sensible heat is acts as a heat source to the atmosphere to raise its temperature.
- ☆ The boundary layer at the bottom of the ocean, which is generally much thinner than the atmospheric boundary layer, is called the benthic boundary layer.
- ☆ Micro-climate is the climate of a small area or climate near ground surface to the top of the plant.
- ☆ Macro-climate is the climate of a large area.
- ☆ Meso-climate is the climate of small area intermediate between microclimate and microclimate such as land-sea breeze and mountain-valley breeze.

- ☆ Eco-climate is a microclimate just above a crop and within the canopy.
- ☆ Phyto-climate is modified form of microclimate produced due to influence of crops.
- ☆ Troposphere is known as weather making layer.
- ☆ Hopkins postulated bioclimatic law in 1938.
- ☆ Isobar is a line connecting various places, having equal atmospheric pressure.
- ☆ Pressure gradient is the differences in atmospheric pressure with horizontal distances.
- ☆ Issallobar is a line connecting places of equal atmospheric pressure gradient.
- ☆ The surface temperature of sun is approximate 6000°K.
- ☆ The nearest distance is 147.5 million km is known as Perihelion and is observed in 3rd January.
- ☆ The Earth rotates west to east direction upon its axis.
- ☆ The average distance between sun and earth is about 150 million km.
- ☆ The longest distance between sun and earth during earth revolution is 152.5 million km is known as Aphelion and it is observed in 4th July.
- ☆ Perigee is the least distance between the moon and the earth.
- ☆ Apogee is the greatest distance between the moon and the earth.
- ☆ Atmosphere is a colourless, tasteless and odourless mixture of gases that surround the earth surface.
- ☆ The mean sea level atmospheric pressure on the earth surface is 1013.25 mb.
- ☆ The mean temperature of earth surface is 15°C.
- ☆ Water vapour comprise up to 4 per cent of the atmosphere by volume.
- ☆ Water vapour is a gas that varies significantly from time to time and place to place.
- ☆ The layer at which chemical composition of gas is uniform is called as homosphere.
- ☆ The layer at which chemical composition of gas is not uniform is called as heterosphere.
- ☆ Troposphere layer is called as "seat of weather phenomena".
- ☆ The Decrease of temperature with increasing height at a mean lapse rate of about 6.5 °C/km or 3.6°F/1000 feet.
- ☆ The Ekman layer exist up to a height of 1000 metres.
- ☆ The temperature increases with increasing height known as temperature inversion.

- Climatology is the science that seeks to describe and explain the nature of climate.
- Meteorology is the physics of the weather phenomenon.
- Agro-climatology is the science that deals with the aspects of climatology having direct relevance to agriculture.
- Agro-meteorology is a branch of meteorology which investigates the relationship of plants and animals to the environment.
- Topo-climate is the climate resulting from terrain characters like altitude, relief, slope *etc.*
- Soil climate is the climate in the vicinity of soil.
- Conduction is the heat transfer process when two bodies of the unequal temperature are in contact.
- Radiation is the process of transmission of energy by electromagnetic waves between two bodies without necessary aid of an intervening material medium.
- Convection is the heat transfer process through movement of particles in fluids and gases.
- The average lapse rate or decrease of temperature with altitude is known as normal lapse rate.
- When there is no change in temperature with altitude, the lapse rate is termed isothermal.
- Adiabatic process is the process of change in temperature without any addition or withdrawal of heat.
- Adiabatic change is the change in the temperature of air due to change in the pressure of the air.
- Adiabatic cooling is an increase in volume accompanied by a decrease in temperature during an adiabatic process.
- Adiabatic warming is a decrease in volume accompanied by an increase in temperature during an adiabatic process.
- The scientific study of earthquake is called seismology.
- Climate variability describes short-term changes in climate caused by naturally occurring events like El Nino or volcanic activity.
- Climate change is a warming or cooling in long-term weather patterns, at different places around the world.
- Crop ecology is the relationship of crop plants to their environment.
- Maximum and minimum thermometer are kept inside the Stevenson screen horizontally.
- Dry bulb and wet bulb thermometer are kept inside the Stevenson screen vertically.

- Diameter of ordinary rain gauge 127 mm.
- Natural syphon rain gauge is commonly used in India.
- Grass minimum thermometer is used to measure the actual minimum temperature experience by the plants near the ground surface.
- The reading of the grass minimum thermometer indicates the possibility of ground Frost.
- When the grass minimum thermometer records 30 °F or below ground frost is likely to have occurred.
- The reading of grass minimum thermometer is noted before sunrise.
- The cup of anemometer moves in an anticlockwise direction.
- Glaze is the ice coating formed when rain or drizzle freezes as it comes in contact with cold objects at the ground.
- Hyetograph is a plot of rainfall intensity against time interval.
- GDD is the departure from the mean daily temperature above the minimum threshold or base temperature.
- Atmospheric window is the long wave radiation escapes to the space between 8.5 and 11 μ. Atmosphere for this spectrum acts as transparent medium instead of absorbing.
- National Bureau of Soil Survey and Land Use Planning is located in Nagpur, Maharashtra and started in 1980.
- National Institute of Disaster Management is located at New Delhi.
- The spectral region of 8.5 and 11μ is used in microwave remote sensing to monitor the features of the sky in overcast sky.
- Counter radiation is a large part of the radiation absorbed by the atmosphere is sent back to the earth's surface. This radiation prevents the earth's surface from excessive cooling at night.
- Sensible heat is the heat that can be measured by a thermometer and thus sensed by humans.
- Latent heat is the energy required to change a substance to a higher state of matter. Change of state through Evaporation and condensation is known as latent heat of evaporation and latent heat of condensation.
- IMD calls storm Super cyclone when the wind blows more than 220 km per hours.
- Prevailing winds is the wind blows more frequently from one direction than from other.
- Paleoclimate is the period of atmospheric change on the order of millions of years.
- Fetch is the length of uniform surface over which the wind has blow.

- InfoCrop model is developed by Indian Agriculture Research Institute (IARI), New Delhi.
- AquaCrop model is developed by Food and Agriculture Organization (FAO).
- WWW stands for World Weather Watch.
- TMC stands for thousand million cubic feet.

References

Adamson, G. (2020). Imperial oscillations: Gilbert Walker and the construction of the Southern Oscillation. In Weather, Climate, and the Geographical Imagination: Placing Atmospheric Knowledges. University of Pittsburgh Press, pp 43-66.

Adedeji, O., Reuben, O. and Olatoye, O. (2014). Global climate change. *Journal of Geo science and Environmental Protection*. 2: 114-122.

Agnidev Manna (2023). Henry Francis Blanford and the Beginning of Scientific Meteorology in India. *International Journal of Humanities Social Science and Management*. 3 (2): 187-192.

Barh, A. and Balakrishnan, M. (2018). Smart phone applications: Role in agri-information dissemination. *Agricultural Reviews*. 39 (1): 82-85.

Bishnoi. O. P. (2010). Applied micro meteorology, applied climatology. Oxford Book Company, Jaipur, India.

Chattopadhyay, N. Ghosh, K. and Chandras, S. V. (2016). Agrometeorological Advisory to assist the farmers in meeting the challenges of extreme weather events. *Mausam*, 67 (1): 277-288.

Chakravartty, N.V. and Sastry P. N. (1985). Some aspects of crop-weather interactions in wheat cultivars. *Int. J. Ecol. Environ. Sci.* 11: 139-144.

Dar, E.A., Brar, A. S. and Yousuf, A. (2018). Growing degree days and heat use efficiency of wheat as influenced by thermal and moisture regimes. *J. Agrometeorol*. 20: 168-70.

Debangshi, U. (2021). Crop microclimate modification to address climate change. *Int. J. of Res. and Review*. 8 (9): 384-395.

https://mausam.imd.gov.in/imd_latest/contents/history.php

https://mausam.imd.gov.in/

https://www.moes.gov.in/programmes/agro-meteorology

https://en.wikipedia.org/wiki/weather_satellite

https://www.weather.gov/about/satellites

https://en.wikipedia.org/wiki/indian_national_satellite_system

https://safar.tropmet.res.in/index.php?menu_id=1

https://mausam.imd.gov.in/responsive/history.php

https://mausam.imd.gov.in/responsive/imdBroucher.php

https://mausam.imd.gov.in/

https://www.tropmet.res.in/

https://www.ncmrwf.gov.in/

https://wmo.int/

IMD (2021): Cyclone warning in India: Standard operation procedure. India Meteorological Department, New Delhi. pp 220.

Katz, R. W. (2002). Sir Gilbert Walker and a connection between El Niño and statistics. *Statistical Science*, 17 (1): 97-112.

Kalsi, S. P. (2005). Satellite Based Weather Forecasting. Satellite Remote Sensing and GIS Applications in Agricultural Meteorology. pp. 331-346.

Kingra, P. K. and Kaur, H. (2017). Microclimatic modification to manage extreme weather vulnerability and climate risks in crop production. *Journal of Agricultural Physics*. 17 (1): 1-15.

Kishtawal, C. M. (2019). Use of satellite observations for weather prediction. *MAUSAM*. 70 (4): 709-724.

Krishnan, A. (1992) Climatic Classiûcation and Agricultural Droughts. In: Crops and Weather (eds. Venkataraman, S and A Krishnan). New Delhi: Publications and Information Division, Indian Council of Agricultural Research, pp.458-508.

Kumar, S., Patel, S. R., Singh, J. and Prasad, D. (2018). Study about different agro meteorological indices under different growing environmental conditions of Raipur district of Chhattisgarh. International *Journal of Chemical Studies*. 6 (4): 2310-2313.

Kumar, M. and Agarwal, L. (2020). Empowering farming community through mobile applications. *Int. J. of Scientific and Technology Research*. 9 (3): 58-61.

Kumar, A., Kumar, M., Pal. V. and Kumar, S. (2018). Study on agrometeorological indices, thermal and photothermal use efficiency of summer groundnut (*arachis hypogaea* L.) at Allahabad Region, India. *Int. J. Curr. Microbiol. App. Sci. 7 (5): 1889-1897.*

Lal, D. S. (2014). Climatology, Sharda Pustak Bhawan, Allahabad.

Mahi, G. S. and Kingra, P. K. (2013). Comprehensive Agrometeorology, Kalyani Publishers.

Mavi, H. S. (1994). Introduction to Agrometeorology. Oxford and IBH Publishing Co. Pvt. Ltd., New Delhi.

Mavi, H. S. and Tupper, G. J. (2005). Agrometeorology principles and applications of climate studies in agriculture. The Hawarth Press, Binghamton, New York.

Mote, B. M and Sahu, D. D. (2014). Principles of Agricultural Meteorology, Scientific Publishers, India.

Murthy, V. R. K. (1996). Basic Principles of Agricultural Meteorology, Shri Venkateswar Publishers, Hyderabad.

Meena, R., Jirli, B., Kanwat, M. and Meena, N. K. (2018). Mobile applications for agriculture and allied sector. Int. J. Curr. Micro. biol. App. Sci. 7 (2): 2317-2326.

Onwuka, B. M. (2016). Effects of soil temperature on some soil properties and plant growth. Journal of Agricultural Sciences. 6 (3): 89-93.

Patil, R. H. (2019). Application of crop simulation models in global agriculture research: A review. Journal of Farm Science. 32 (4): 1-11.

Pathak, H., Mishra, J. P. and Mohapatra, T. (2022). Indian Agriculture after Independence. Indian Council of Agricultural Research, New Delhi 110 001, pp 426.

Premdeep, R. N., Khichar, M. L. and Sharma, A. (2017). Quantification of Energy Indices Requirement of Cotton Varieties under Different Growing Environments. Int. J. Curr. Microbiol. App. Sci. 6 (3): 809-817.

Rao, V. U., Rao, A. V., Kumar, P., Bapuji Rao, B. and Sastry, P. S. (2013). Agro meteorological Aspects of Extreme Weather Events, Central Research Institute for Dryland Agriculture, Santoshnagar, Hyderabad, 303 pp.

Raghunath, H. M. (2006). Hydrology. New Age International (P) Ltd., 18-21.

Rana, G. and Yadav, M. K. (2014). The Sun and its activities. Int. J. of Engg. Res. and General Science. 2 (6): 784-791.

Singh, V. P. (1994). Elementary Hydrology. Prentice Hall of India Private Limited, 174-175.

Siddhartha, K. (2013). Atmosphere, Weather and Climate, Kisalaya Publications Pvt. Limited, New Delhi, India.

Singh, M., Niwas, R. Godara, A. K. and Khichar, M. L. (2015). Pheno-thermal response of plum genotypes in semi-arid region of Haryana. Journal of Agrometeorology. 17(2): 230-233.

Sarkar, S., Kumar, B. and Kumar, S. (2021). Mobile applications for Indian agriculture and allied sector: An Extended Arm for Farmers. Int. J. Curr. Microbiol. App. Sci. 10 (3): 1913-1920.

Singh, M. and Bhatia, H. (2011). Thermal time requirements for phenophases of apple genotypes in Kullu valley. J. Agrometerol. 13 (1): 46-49.

Sastry, P. S. N., Charkravarty, N. V. K. and Rajput, R. P. (1985). Suggested index for characterization of crop response to thermal environment. Int. J. Ecol. Environ. Sci. 11: 25-30.

Thornthwaite, C. W. and Mather, J. R. (1955). The water balance, Publications in climatology. 8 (1), Drexel Institute of Technology, Laboratory of Climatology, Centerton.

Tripathi, A., Fischer, M., Klem, K. and Marek, M. (2018). The evaluation of radiation uses efficiency and leaf area index development for the estimation of biomass accumulation in short rotation poplar and annual field crops. Forests. 9: 168-178.

Thirupathi, M. Shashikala, A., Prabhakar, M. (2015). Variability study on the length of growing period. International Journal of Advancement in Remote Sensing, GIS and Geography. 3 (2): 48-58.

Venkatasubramanian, K., Tall, A., Hansen, J., Aggarwal, P. (2014). Assessment of India's integrated agrometeorological advisory service from a farmer perspective. CCAFS Working paper no. 54. Copenhagen, Denmark: CCAFS.

Index

R

S

T

U

V

W